D1066032

Sustainable Development and Innovation in the Energy Sector

Ulrich Steger · Wouter Achterberg (†) · Kornelis Blok
Henning Bode · Walter Frenz · Corinna Gather
Gerd Hanekamp · Dieter Imboden · Matthias Jahnke
Michael Kost · Rudi Kurz · Hans G. Nutzinger
Thomas Ziesemer

Sustainable Development and Innovation in the Energy Sector

With 33 Figures

 Springer

b 2594034X

For the Authors:
Professor Dr. Ulrich Steger
IMD Lausanne
P.O. Box 915, 1001 Lausanne
Switzerland

Editing:
Friederike Wütscher
Europäische Akademie GmbH
Wilhelmstraße 56, 53474 Bad Neuenahr-Ahrweiler
Germany

Library of Congress Control Number: 2004111215

The German edition of this book was published under the title „Nachhaltige Entwicklung und Innovation im Energiebereich" by Ulrich Steger et al, © Springer-Verlag Berlin Heidelberg New York 2002

ISBN 3-540-23103-X **Springer Berlin Heidelberg New York**

This work is subject to copyright. All rights are reserved, whether the whole or part of the material is concerned, specifically the rights of translation, reprinting, reuse of illustrations, recitations, broadcasting, reproduction on microfilm or in any other way, and storage in data banks. Duplication of this publication or parts thereof is permitted only under the provisions of the German Copyright Law of September 9, 1965, in its current version, and permission for use must always be obtained from Springer. Violations are liable to prosecution under the German Copyright Law.

Springer is a part of Springer Science+Business Media
springeronline.com
© Springer-Verlag Berlin Heidelberg 2005
Printed in Germany

The use of general descriptive names, registered names, trademarks, etc. in this publication does not imply, even in the absence of a specific statement, that such names are exempt from the relevant protective laws and regulations and therefore free for general use.

Cover design: Erich Kirchner
Production: Luisa Tonarelli
Typesetting: Köllen Druck + Verlag GmbH, Bonn + Berlin
Printing: Mercedes Druck, Berlin
Binding: Stein + Lehmann, Berlin

Printed on acid-free paper   30/3141/LT – 5 4 3 2 1 0

HC
79
.E5
S8649
2005

Preface to the Translation

The *Europäische Akademie zur Erforschung von Folgen wissenschaftlich-technischer Entwicklungen Bad Neuenahr-Ahrweiler GmbH* is concerned with the scientific study of the consequences of scientific and technological advance both for the individual and social life and for the natural environment. The Europäische Akademie intends to contribute to a rational way of society of dealing with the consequences of scientific and technological developments. This aim is mainly realised in the development of recommendations for options to act from the point of view of a long-term societal acceptance. The work of the Europäische Akademie mostly is carried out in temporary interdisciplinary project groups whose members are notable scientists from various European universities. Overarching issues, e.g. from the fields of Technology Assessment or Ethics of Science, are dealt with by the staff of the Europäische Akademie.

The results of the work of the Europäische Akademie is published in the series "Wissenschaftsethik und Technikfolgenbeurteilung" (Ethics of Science and Technology Assessment), Springer Verlag. The academy's study report 'Nachhaltige Entwicklung und Innovation im Energiebereich' was published in October 2002. It contains a straightforward strategy how innovations can help to achieve a sustainable development in the energy sector. The academy decided to provide for an English translation of this report that is published in the present volume in order to make this strategy available to a wider scope of recipients.

Bad Neuenahr-Ahrweiler, June 2004   Carl Friedrich Gethmann

Preface

In discussions concerning sustainable development, innovations are often cited as a "miracle cure". Through innovations, we are to prevent a situation where an increase in output leads to an increase in the consumption of natural resources. This means for the of energy sector: Innovations should help us to reconcile the further growth of the national products of the industrial countries, and at?? the backlog demand of the developing and emerging nations, with a reduction in the consumption of non-renewable energy resources, which must not give rise, however, to an inappropriate consumption of other resources.

The core question addressed by the interdisciplinary project group, "Sustainable development and innovation in the energy sector", which was established by the Europäische Akademie (european academy) in September 2000, was therefore: "To what extent can innovations lead to a sustainable energy system?" The members of the group were selected according to their competences within their disciplines with regard to the subject to be dealt with. The project time frame was 20 months, of which 13 days were spent in plenary session.

The final report presented here derives from chapters, which were drafted, under the direction of one of the group members, by individual working groups before being integrated by the plenum. The work of the project group was based on the judgment that "interdisciplinary research" does not exist as such, but disciplinary competences are a prerequisite for dealing with individual aspects of the subject. An integration of the various disciplinary perspectives, methodologies and results with regard to the non-disciplinary question at hand follows as the second step. The procedure pursued by the group was transdisciplinary, in this sense. The result is a text that is consistent in itself, and a coherent argumentation that can be examined step by step (even if the disciplinary background of the "original author" is easily detected in some sections of the report).

The group was open to continuous inspection by external specialists. The work schedule was discussed at the kick-off workshop in January 2001.

We thank our colleagues, Professor Dr. Wilhelm Althammer (Handelshochschule Leipzig), Professor Dr. Nicholas Ashford (MIT), Dr. Gerd Eisenbeiß (Forschungszentrum Jülich), Dr. Klaus Rennings (ZEW Mannheim), Dr. Herwig Unnerstall (Universität Leipzig), Professor Dr. Alfred Voß (Universität Stuttgart) and Professor Dr. C.-J. Winter (Energon) for their valuable suggestions and pointed criticism, which both helped to provide a precise orientation for this study. At the mid-term workshop in November 2001, a first draft was presented to the following colleagues: Professor Dr. Dr. Brigitte Falkenburg (Universität Dortmund), Professor Dr. Wilhelm Althammer (Handelshochschule Leipzig), Dr. Gerd Eisenbeiß (Forschungszentrum Jülich), PD Dr. Volker Radke (Berufsakademie Ravensburg),

Dr. Klaus Rennings (ZEW, Mannheim) and Dr. Herwig Unnerstall (Umweltforschungszentrum Leipzig). We also extend our thanks to the participants of that meeting, for their meticulous comments, which later helped to round off the study. Thanks to a good working discipline, the materials on which the discussions were based were ready in time for almost every session. The intellectually stimulating working atmosphere, characterized by professional respect and friendly cooperation, allowed for intensive, constructive, at times controversial discussions and mutual learning throughout various perspectives and methods.

The group's productivity was fostered, not least, by the hospitality of the Ahr valley, and the friendly and efficient services, with which the Academy staff supported our work, especially Ms. Pauels, to whom we would like to express our gratitude. We also thank Mr. Jochen Markard and Mr. Joachim Schmidt-Bisewski, who accompanied the project through the early stages, as well as Ms. Sevim Kilic of the european academy, who worked on the text and prepared it for publication.

Lausanne, June 2002

Table of Contents

Summary

Introduction

The discussion seems to be paradox: Almost every energy scenario is based on trends that would lead to an enormous growth in the demand for energy in the coming decades. Meanwhile, at international conferences, among other places, one is concerned with the opposite outlook, a massive reduction of greenhouse gas emissions, especially of CO_2 emissions caused by energy consumption. Experts also point out the political risk of depending on mineral oil and remind us of the fact that resources are not inexhaustible. How can this chasm be overcome? How can we build a more sustainable energy system from the existing one? Hopes are mostly pinned on technological progress and innovations.

So far, however, there are no specific suggestions concerning the extent to which innovations can really contribute to reconciling ever-growing energy consumption with the limitations regarding the availability of resources and the environment, as well as with the structural demands on any energy system.

The aim of this study is to bring together economic, legal, scientific and philosophical competencies with a view to developing such proposals. This task requires clear focusing on the intersection of the three central issues, i.e. energy, sustainable development and innovation. A comprehensive treatment of the three subject fields was not intended. Neither could many of the debates in this context be dealt with beyond their relevance for the strategy proposal of this study.

In deriving our recommendations, the aims laid down by democratically legitimized agencies were taken into account, no matter how vague these aims are, especially on the international level. An important part of our work concerned the analysis of conflicting objectives in economic policy and the question of how such conflicts can be overcome through a more comprehensive, incentive-based mix of instruments tailored to the specific substance of an innovation.

Terminological and conceptional foundations

Since a sound investigation cannot be performed without a clearly defined terminological and conceptional framework, we will start by inspecting the central concepts of sustainability, energy and innovation.

The idea of sustainability with its two normative cornerstones of intra- and intergenerational justice has to be made concrete especially for the area of energy which is based mostly on exhaustible energy sources. Instead of a static concept of stocks, which conceptually excludes a sustainable use of limited resources, a dynamic concept of flows (current use) is introduced, which is based on the substitution of non-renewable resources by renewable ones and on the continuous creation of new, more efficient ways of using resources. In this way, the need for innovations in this area is, at least implicitly, addressed. If, by appropriate innovations, one succeeds in reducing, the use of exhaustible resources in production and consumption, so that a lower consumption of such limited stocks will suffice in the future, the chances to utilize such declining resources can be maintained or even improved in some cases. The *possibility* of such chances, however, does not imply that, faced with the present trends in the areas of energy use, strains on the environment, private consumption and population development, a path of "sustainable development" can *actually* be found.

For the sake of clarity, our analysis distinguishes between *sustainability* and *sustainable development:* The regulative idea of sustainability initiates and accompanies, with a practical intention, a search and learning process which leads to the more concrete concept of sustainable development, whereby potentials and possibilities for action towards sustainability can be identified; hence sustainable development is regarded, in principle, as a guide for action.

Considering the multitude of efforts to define "sustainable development" – by now, there are more than 200 of them after fifteen years of scientific and political discussions –, one cannot but admit that this concept is still very vague or, sometimes, even mired in confusion. In the present discussion of the problems surrounding sustainability, a first approach leads to the observation of three different ways of dealing with the varied meanings of "sustainable development": Apart from sheer *disapproval* (because of the "woolliness" of the concept) and an *integrative strategy* (by burdening the concept with everything that happens to suit one's purposes), there is another possible attitude, which is shared by our group: the effort to deal with the concept in a *productive* manner and to define it as precisely as possible according to scientific criteria. This involves comparing various possible definitions of the concept and asking the question which concrete conclusions follow for the central research question of our investigation for each case. This path is taken in neoclassical environmental economics on the one hand and on the other hand especially by ecological economics, the "science of sustainability". One has to find a balance between overdetermination and underdetermination of this concept and one should neither burden it with specific requirements which meet the most stringent ecological criteria, but make it an unachievable ideal, nor should one leave it so vague that it can mean everything and effect nothing: In principle, sustainable development must be an operational concept.

The various concepts ranging from "weak" to "very strong" sustainability differ with regard to assumptions about substitution and complementarity between man-made and natural capital. This study applies the concept of critical sustainability based on a concept of critical natural capital, taking into account few, but crucial and hence critical "crash barriers" or "bottlenecks". Our interpretation of sustainability thus is related to the far-advanced discussion of setting environmental standards.

Energy may determine our everyday life and constitute an important production factor in economic theory; from the physics point of view, however, it is a rather abstract entity, which can only be defined accurately in terms of a differentiated mathematical model. Historically, the concept of energy was initially defined simply as the "potential to perform work". In that sense, of course, energy is not conserved; this is why the notion of "energy *consumption"* has become common usage.

The connection between the, at first, entirely different concepts of "energy" and "heat" was clarified only in the 19th century, with the formulation of the First Law of Thermodynamics stating that energy is preserved, i.e. it is neither created nor destroyed, but just transformed from one form into another. (At the beginning of the 20th century, the concept of energy was extended by Einstein in his theory of special relativity, which includes mass as a form of energy.) Hence, energy consumption actually means energy *degradation* i.e. transforming valuable or available energy (exergy) into lower-value or non-available energy (anergy). The boundary between exergy and anergy is not absolute, but depends on the system considered. For instance, water at a temperature of 20 degree Celsius in an environment at zero degree contains usable energy (exergy), while this would not be the case at an ambient temperature of 20 degree.

The *energy system* (of a country or the Earth as a whole) is defined as the overall structure of the primary energy resources being used, the infrastructure for their distribution and transformation into final energy and the specific demand structure of so-called energy services. With regard to the quality of the energy, the distinction between the demand for heat or work, respectively, plays a special role, as well as the differentiation between stationary and mobile demand and the function of electricity. Together the supply and demand structures determine the potential for changing an existing given energy system.

The term innovation describes a new problem solution prevailing in the market, in connection with new factor combinations. Sustainable innovation means factor combinations and new problem solutions that lead to less environmental strain and a reduced consumption of resources, without necessitating restrictions on other social objectives. An innovation does not have to be a new technological solution; it can also be a new service or a new form of organization.

In order to invigorate sustainable innovation, one requires knowledge on innovation determinants. The extent, the direction and the speed of innovation activity in a national economy depend on a multitude of factors, which are sometimes summarized as the "national innovation system"; these reach far beyond research and developments politics, touching on tax and education systems. In the course of European integration, it has become more appropriate in some areas to speak of a European innovation system. This entire context needs reshaping, if innovation activity is to aim at a sparing use of resources. For policies concerning innovations a double strategy appears to be called for, which, on the one hand, aims at short-term effects while, on the other, providing longer-term direction.

Through general improvements of the framework for sustainable innovation activity (e.g. regulation reform, tax reform, basic research priorities), the search efforts of scientists and inventors are steered into a different direction; the common pool of knowledge and ideas (the pool of inventions) is enriched accordingly. This part of the double strategy requires more time and has a general increase of sustainable innovation activity as its objective, rather than sector-specific potentials or specific types of innovation.

These components complement each other. Successful innovation policies emerge from the well-adjusted combination of both. As the transitions between the two kinds of innovation policies are fluid, arguments (partly shaped by ideology) about which one to choose will be unfruitful. Sustainable innovation policies as a whole have the objective to change the framework in such a way that the chances of sustainable innovation potentials to prevail in the market are improved. The advance of the framework for sustainable innovations finds itself confronted with the problem that successes are the result of long-term developments and cannot be causally attributed to certain changes of individual conditions. Hence, the political acceptance of reform especially of this kind is difficult to achieve.

From the factors of success and the obstacles thus identified, political recommendations can now be derived, where – after the acknowledgement of the principal need for innovation policies by the state – the choice of the concrete technology, the instruments, their dosage and the phase specifics are the main concerns. The recommendation emerging is a well-dosed combination of sponsorship for research far from the market (fundamental research), a search and discover function through competition, followed by support for an accelerated diffusion, and a general improvement of the conditions for sustainable innovation activity.

Normative evaluation and decision criteria

The application of traditional rules of decision-making requires a precise formulation of the relevant options for action and the environmental conditions influencing the effects of the action. However, for those cases where it is not clear which environmental conditions must be taken into consideration (decisions under profound uncertainty), these rules cannot be applied. Since long-term environmental transformations are characterized by profound uncertainty, different techniques of decision making have to be applied.

In such cases, environmental politics calls upon the "precautionary principle", according to which preventive measures are permitted even if the scientific evidence is not conclusive, but merely plausible. The costs of such measures must be proportional (principle of proportionality), preferably, due to the profound uncertainty involved, in comparison to another end easier to specify. Precautionary measures with regard to climate change, for instance, can be assessed through the ends of secure supplies and a reliable energy system.

A set of ends for a sustainable reorganization of the energy system is listed in the following table:

[Table 3.1] Objectives for a sustainable restructuring of the energy system

Objectives	Concretization
Availability of resources	Period of secure practice
Energy system	Reliability (end user), openness of options, risk avoidance
Environment	Climate change, emissions, surface consumption

Whenever in this study we point out the necessity of a sustainable restructuring of the energy system, one has to keep in mind all of the ends cited above. However, for this set of ends, the reduction of CO_2 emissions can serve as a "guiding indicator" that is supplemented by further indicators (surface consumption, openness of options etc.) in a particular context.

Every concept of sustainability involves certain normative decisions with specific ethical implications. The position that there is no such thing as an obligation towards future generations is – as far as we can see – hardly ever advocated as such. However, it does come out in the argument that the interests of future generations can be taken account of today only insofar as the present generation is not harmed (intertemporal pareto-improvements). This "win-win" concept may suit innovations, because the accumulation of technological and organizational knowledge may compensate future generations for the diminished resources that will be at their disposition.

The obvious problem with this position is, however, that as soon as the margins for pareto-improvements close in and an actual trade-off situation and hence a grave ethical conflict arises, conclusions become impossible. Therefore, employing an intertemporal pareto criterion can only be a first, largely ethics-abstinent step towards answering the questions we are concerned with. Are intertemporal pareto-improvements not possible in a particular case, one must look for justifications for the trade-offs (balances) between different options.

The question whether issues of intergenerational distribution – which are pertinent to the debate on sustainability – offer any leeway for pareto-improvements, as in the case of the sustainable energy innovations discussed in this paper, or if we face a trade-off situation at least in some areas, is of course an empirical one, which has to be answered for every individual case. However, any conception of sustainability introduces some type of intertemporal stock. The demand to conserve it – or the assumption that this would be fair – puts the concept of sustainability into a normative context.

Long-term responsibility is a fundamental aspect of the concept of sustainable development, which can be – apart from its form in detail – assumed unproblematic through recourse to a robust, moral intuition. Everybody will accept long-term obligations at least towards the generations immediately subsequent. There will also be an intuitive acceptance of the binding character of this obligation fading with the distance in time *(gradation)*, for one will rather afford one's children a certain advantage, or spare them some harm, than one's descendents in the tenth generation. This gradation of the binding character can also be justified by the increasing uncertainty of the occurrence of the desired effects of action.

A notion of intergenerational justice is woven into the concept of critical sustainability, in the sense that the standards regarded as critical shall be adhered to. The concept of critical sustainability recurs to issues of gradation, most importantly in terms of the uncertainty of the relevant knowledge, as discussed above.

Naturally, we cannot take into account future generations and their needs *correctly*, since these do not exist yet. For precisely this reason, we often see a distortion of the balance between ecological, social and economic criteria within the concept of sustainability: While the advocates of social and economic aspects rely on special interest groups, ecological aspects are only supported by environmental groups or agencies. The latter consist of members of the generations living today, but acting in favor of the supposed interests of future generations. At this point, science comes into play, for science cannot speak as an advocate for future generations. It should help, however, to make transparent the risks of overstretching ecological resources, by using the best scientific results available. Therefore, we have to describe risks and future developments with uncertain effects, instead of providing simple "recipes" made up from clear facts.

Towards a sustainable energy system – deficits and points of reference

The idea of *sustainability as a legal standard* is relatively new. It was first taken up in the realms of international law. In the documents that emerged from the "Earth Summit" in Rio de Janeiro in 1992, this notion is found especially in Agenda 21. It was made legally binding, somewhat later, through specific treaties, namely the *United Nations Framework Convention on Climate Change*, which came into force following its ratification by 160 countries and was concretized in the *Kyoto Protocol* of 1997. The difference of interests between developed countries and developing countries as well as the diverging ideas about the relative weight of economic and ecologic interests in the industrialized countries have stripped this agreement of most of its effectiveness over recent years.

On the European level, since the Amsterdam renegotiations and already in the preamble of the original EU treaty, sustainable development is cited as one of the objectives. In the German constitution *(Grundgesetz)* (Art. 20a), "protecting the natural foundations of life" is laid down also with regard to the responsibility towards future generations.

However, the sustainability of the energy system must not be analyzed exclusively from the aspect of climate protection. At the foundation of the International Energy Agency (IEA) following the oil crisis of 1973/74, the *security of supply* (procurement) was the principal consideration. Today, about 50% of the energy demands of the European Union are covered by imports. In geopolitical terms, ca. 45% of the oil imports origin from the Middle East; 40% of the natural gas imports stem from Russia. By 2020 – according to a EU prognosis – the import component will have risen to 70% again; over the same period, we will see a shift to a renewed dependence on the Middle East, where about two thirds of future oil reserves are located and where, as estimated by the IEA, more than 85% of any additional production capacity is likely to be found.

Presently, neither the international nor the national legal standards are precise enough to derive direct, operative "sustainability targets" from them. Effective political action, however, requires precisely formulated objectives and the corresponding knowledge on action, both of which can only be developed in dialogue with the sciences. In particular, the sciences have to *analyze* the present energy systems and to formulate a precise *benchmark* for a sustainable provision of energy. The cornerstones of this analysis are the assertions that (1) commercial energy consumption has risen by a factor of 5 over the last 50 years, (2) more than 90% of this energy stems from fossil resources and (3) differences between the poorest and the richest countries, concerning the availability of commercial energy, are more than hundredfold. Most prognoses predict that the global energy demand will double or even grow by the factor of four until 2050. In contrast, for reasons of climate protection and supply security the consumption of fossil energy should be cut by half over the same period.

Two recepies are usually invoked to close the gap between demand and critical limits: *enhancing energy efficiency* and *decarbonization*. The first aims at decoupling the gross domestic product (GDP) from the energy demand, the latter at the substitution of fossil resources by renewable or carbon-free energy sources. The present development of the global energy system shows that both processes are far too slow to stop the growth of CO_2 emissions, let alone to reduce them. In other words, the present development of the global energy system is not sustainable, neither with regard to climate protection nor from the perspective of energy supply security, especially if the geographic distribution of the fossil energy resources is taken into account.

The essence of the above considerations can be summarized by the following simple calculation: Taking into account the typical growth target of the GDP of 2% for the EU and other industrialized countries, and assuming a target of 2% for the reduction of the CO_2 emissions, we need a decrease of the CO_2 intensity (CO_2 emissions per GDP) of 4% per year. The CO_2 intensity is used as a guiding indicator for a sustainable transformation of the energy system.

In the following evaluation, the scenario "S450" of the Intergovernmental Panel on Climate Change (IPCC), which appears to be ambitious without being unrealistic, plays a central role. In addition, long-term trends, the need for political stability in the north-south relationship as well as a reduction of the dependence from Middle East oil are also important goals. To underline the last point we mention that at present the largest consumer of oil, the US, imports about 40% of its oil demand.

In order to operationalize the sustainability targets, two concepts are introduced: (1) The *time of safe practice* is based on the idea to characterize societal activities by the (hypothetical) time during which that activity could be carried on until it reached its limits (e.g. because of resource limitations, environmental strain etc.). (2) The *inertness* of the energy system can be defined as the time needed for a significant change of the system. For the present energy system, such a significant change could result in, for instance, the complete substitution of the present fossil energy supply by renewable energy resources.

With the help of these concepts, the aim of sustainability can be defined as follows:

(1)   A practice (e. g. an energy practice) is sustainable if the time of safe practice is constant or growing (principle of the constant time of safe practice).

(2)   The time of safe practice must exceed the inertness of the system concerned.

Applied to energy, this means that a sustainable energy system has to be supported by two pillars: (1) the efficient use of energy and (2) a growing use of solar and other renewable resources. With the technologies known today, the present standard of living in the EU could be maintained with an energy consumption of *2,000 Watt per person*. (The present demand in the industrialized countries amounts to between 4,000 and 10,000 Watt per person). This demand could be met in a sustainable manner, i.e. largely by renewable resources. The *2,000-Watt benchmark* forms the basis for our further considerations. We have reasons to assume that there is enough time for such a transition, provided that the process is vigorously initiated now.

Potentials and barriers for a sustainable energy system

The potential of innovations in the energy sector presently in development or recently introduced to the market is fundamental for such a transition. Hence, their assessment at different stages of development, or at the early stages of commercial exploitation, is the next step of the investigation towards a sound evaluation of their potential concerning energy efficiency.

There have been periods in the past – for instance periods of high energy prices – when the energy efficiency of new appliances improved by at least 5% annually. Now we pose the question if such an improvement can be achieved in the future too. By citing representative examples (energy consumption for residential heating, cars, electricity production and selected industrial processes), we show that the technical means are actually in place for realizing an annual improvement of upwards of 5% in the future. Already today, we are even able to specify precisely the potential for the coming, say, 15 years.

An improvement of the energy efficiency of new appliances of, annually, about 5% makes possible to halve, compared to the present situation, the total demand for energy by the year 2050. This calculation takes into account further economic growth and a slow turnover of capital stocks. Such a reduction in the energy demand is a precondition for a decisive growth of the relative contribution of renewable energies.

The present contribution of renewable energies in the energy market exceeds the general expectation of 20 years ago by a considerable margin. Wind energy will become commercially viable over the coming years. Photovoltaic energy still has a long way to go before achieving that goal. The wide spectrum of applications of biomass through a variety of technologies will become viable on a timescale somewhere in between. Short-term applications include its burning large power plants and its fermentation in small plants. A plethora of new technologies based on gasification for producing gaseous (for instance for the combined production of heat and electricity) and liquid fuels (e.g. for the transport sector) offer promising prospects.

Taking into account the different types of renewable energy sources, we developed different scenarios for meeting at least half of the energy demand by using renewable energy sources.

0. The reference case: Continuing existing trends such as a slow improvement of energy efficiency, a gradual rise in the final-energy demand, an increasing contribution from natural gas, the phasing out of nuclear energy and small contributions from regenerative energy sources
I. A scenario involving the energy demand being relatively stable (albeit with a demand shift from heat to electricity) and a supply system based on the – under the condition of limiting the CO_2 emissions – cheapest abundant energy sources i.e. biomass and natural gas
II. The same demand development, but less use of biomass
III. A scenario with a markedly lower energy demand

Hydropower still is the predominant renewable source for electricity production at present. For the coming decades, the largest growth is expected in the areas of wind energy and biomass. The prospects for biomass are most favorable, though the

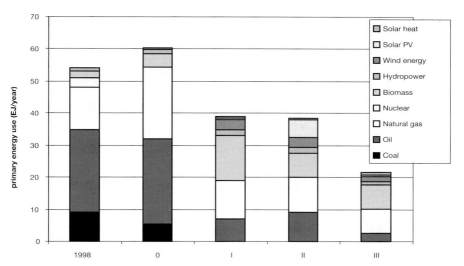

[**Fig. 5.5**] Overview of the primary energy inputs in the year 2050 for the three images.

surface demand involved will be enormous. Still, even within the densely populated European Union an extension of biomass exploitation sufficient for the biomass intensive scenario above (scenario I) is possible. Photovoltaic energy can only become important in the longer term. It is conceivable for this source to play a major role, but that would require extensive development efforts and cause higher costs. Hence, there are several combinations of energy efficiency improvements and renewable energy sources that can reduce the demand for fossil energy in the European Union by 60 to 80%, compared to the present status, by the year 2050.

The developments lined out here will never happen without targeted measures. There are various obstacles to overcome. In many sectors, energy does not represent an important cost factor. This is definitely true for the service and agricultural sectors. Even in residential households, the energy costs are spread over several areas (mobility, heating, electricity etc.). Therefore, the energy costs are often not considered appropriately when making decisions. Furthermore, the positive external effects (e.g. fewer emissions) are not appreciated properly in the market. New technologies reside at the upper end of the learning curve and cannot compete easily with established technologies, which have been optimized over a long period of use. On the other hand, the advantages of mass production are crucial if energy efficiency and renewable energies – especially if one deals with manufactured technologies – are to be competitive against established on-site technologies. In many cases, new technologies have to be compatible with existing plants and comply with existing standards and infrastructures, which again delays market penetration. This applies especially to a capital-intensive sector where „sunk costs" prevent a rapid turnover of capital.

Apart from that, the history of substituting one energy source by another shows that new energy sources must be not just competitive, they also have to offer additional advantages (for instance the "cleanness" of oil compared to coal). The substitution process is never untouched by politics (in every direction), and it strongly depends on the service life of the existing energy infrastructure. However, as soon as new technologies have acquired a critical mass, the substitution accelerates.

The reality of sustainability: Conflicts of aims in the choice of instruments

The promotion of "welfare" or the "public weal" is often cited as the purpose of political or ecopolitical action. The meaning of this is, however, often unclear when one deals with concrete measures. The reason is that certain measures appear advantageous in some respect while, at the same, they often imply drawbacks too. Such drawbacks can be the uneven distribution of the costs and benefits of a measure or that the removal of one problem implies the creation of a new one.

A well-known example is the magic quadrangle in macroeconomic policy: According to the law of stability and growth, a high level of employment, low inflation, external trade balance and appropriate growth have to be the objectives. In fact, however, policies also aim at a fair distribution. These are five *aims of economic policy*, with the effect that measures for improving the achievement of one aim can easily compromise the realization of one or several other aims. If, for example, a higher level of employment is achieved, this can create a risk of higher inflation and more imports, and possibly leads to lower achievement concerning two of the aims, namely „low inflation" and "trade balance". If one aims to achieve a fairer distribution the wage rises, this can lead to less employment and growth. Again, two aims are met less successfully, if one promotes another aim. Hence we are dealing with *conflicts of* aims.

For this study, the conflicts between environmental aims and other ones are of particular importance. Based on the "new microeconomics", we discuss conflicts rooted in market shortcomings and distribution problems, where employment, competition, trade, finance and development policies are relevant. A reduction of environmental emissions, especially in the energy sector, requires the fall in production and thus either in employment or in wages, if employment levels are to be maintained.

If cost increases due to environmental policy measures cause a fall in production volumes on the factory level, the result is the same as that of monopolist action, where the monopolist sets a monopoly price. Thus, such measures can intensify monopoly effects, if they do not consider them properly right from the beginning.

A large part of energy emissions stems from (international) transport. (International) trade serves to enable the consumer or enterprises to buy goods at more favorable prices. If transport costs rise for environmental reason, so that the environmental costs are borne by the polluter, this runs against the interests of transporters, importers and receivers of goods.

The weakness of environmental policies limited by national or regional boundaries lies in the fact that enterprises can migrate to other regions where they do not have to bear environmental costs. Empirically, this effect may well be minor because international competitiveness was given precedence, often through regulations for exceptional cases. In the absence of such regulations, however, the effect partially leads to a deteriorating of employment levels and subverts the environmental policy itself, for instance because emissions then come in from less regulated countries.

To make the costs of environmental policies efficient, one is looking for ways to employ funds from industrialized countries in developing countries, if a stronger effect can be achieved there. If this leads to a stronger demand for land, for instance to realize reforesting programs within the framework of the *Clean Development Mechanism* (CDM) agreed in the Kyoto Protocol, it can mean higher rents asked from small tenant farms and higher food prices, which is in obvious conflict with the development aim of reducing poverty.

Conflicts of aims, as long as they exist, can forestall political decisions, because individuals, especially politicians and lobbyists, can differ in their judgment of the importance of different aims. In particular, they can differ in their acceptance (or non-acceptance) of the actual existence of a problem or in their assessment of the extent of a problem. It can be very expensive in the long term, when no measures are taken because there is insufficient information even if a relevant problem is indeed very important, or when measures are decided which later turn out to be unnecessary.

Consequently, the conflict-laden distribution effects of environmental efficiency gains have to be defused in order to reduce resistance. This can be achieved through innovation-political measures. The promotion of superior technologies – in terms of their environmental effects – can cut the marginal costs of enterprises, increase employment and reduce monopoly prices as well as transport emissions without hampering international trade or causing capital migration. Such technology support takes place at home, not in developing countries. The import of superior technologies only reduces emissions if these technologies become the standard and older technologies disappear from the market. To that extent, the contrasting interests outlined above are absent in the employment of innovation policies. Innovation policies can be used as a complement to other environmental measures. In order to gain approval for measures such as environment taxes and certificates, one can offer innovation incentives softening the effects of the cost distribution.

Within the legal framework of the European Union, rules have emerged on what must be considered when economic aims – for instance the free movement of goods and services in the internal market – are put aside in favor of the protection of the environment. The reasons must be compelling and the instruments used must affect the internal markets as little as possible. The rules often require only a temporary intervention or the setting of threshold values (e.g. in the case of subsidies for environmental protection technologies). The European Court of Justice has laid down strict rules concerning the corresponding evidence required.

Conflicts with EU competition laws can arise in two respects. State incentives have to be measured against the prohibition of subsidies, which covers all national incentive measures favoring the recipient financially, but, in the view of the European Court, does not affect the rules concerning purchase volumes and compensation duties, which mainly present a financial burden for private-sector energy suppliers and lead only indirectly, if at all, to national revenue losses. Any existing subsidy can be justified on environmental reasons, if it is in accordance with the common framework for state subsidies in the environmental sector. This framework provides special rules for regenerative energies, allowing, in principle, at least temporary subsidies. The polluter-pays principle, which dominates the common subsidies framework too, gives reason for concern.

The second source of conflict in terms of competition law exists in the competition rules governing enterprises. Where self-regulation within the private sector, e.g. concerning CO_2 reduction or regenerative energies, leads to cooperation between companies and thus affects free competition, this too can be justified, if it is inalienable, on reasons of environmental protection. Insofar as it offers the prospect of similar effects as state support measures or market interventions, such self-regulation can also put in question the necessity of restrictions to the movement of goods.

Strategies to accelerate energy innovations

A strategy has to activate the potentials with regard to the 2,000-Watt benchmark without failing at the obstacle of conflicting aims, i.e. the balancing of the three pillars of sustainability. The analysis above allows developing such a strategy with a bundle of measures to promote sustainable energy innovations:

1. We propose a strategy to accelerate sustainable energy innovations through custom-made support measures for different phases of their life cycle within a learning curve model. At the beginning of the life cycle, subsidies should help achieving the cost advantages of the effects of scale by enabling enterprises to move faster along the "learning curve" of cost reductions. In a later phase, measures of self-regulation as negotiated solutions or as unilateral self-commitments of the private sector should lead to a faster penetration of the market.

   The focus is on energy-efficient technologies at the stage of their market introduction, meaning that pilot projects and demonstration proposals already exist and the technology concerned is now at a stage where an "early adapter" has to be found and industrial production and service structures to be developed through higher quantities. In the majority of cases, these subsidies take the shape

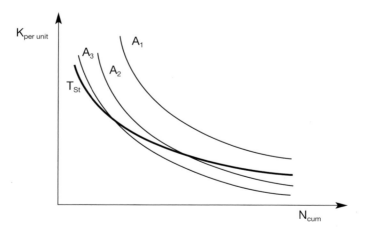

[Fig. 7.1] Learning curves for energy-efficient technologies compared to the standard technology (T_{St}). ($K_{per\ unit}$: Unit costs, N_{cum}: accumulated production volume)

of start-up financing. For it can be shown that, in most cases, the costs of the clean technology do not exceed those of the old technology in the long run (see curves A2 and A3 in the graph below). In the face of energy markets characterized by deregulation and volatile prices, such support is particularly important for regenerative energy sources. The different technologies do not all start from the same situation. Energy production from wind power is much closer to profitability than photovoltaic energy. Energy from biomass could turn out too expensive in the industrialized countries of the northern hemisphere, but not in developing tropical countries. Therefore, the promotion of various technologies must be customized depending on their position on the learning curve and the answer to the question, how soon can they profit from economies of scale. For the decentralized, "manufactured" technology of energy production from regenerative sources, the efficient mass production is the most certain way to compete with "on-site" technologies such as power.

For this approach to subsidies, the Dutch model of the "energy list" is referred to specifically. By maintaining a list of technologies eligible for support, which is updated annually, one avoids subsidizing a technology for a longer period than necessary. The experience gained in the Netherlands – experience with overcoming information asymmetries, minimizing free-rider effects and focusing exclusively on technologies that are in fact innovative – have to be taken into consideration.

2. Demand should be further stimulated through state procurement programs. For instance, in the course of regular construction, modernizing and repair measures over the next 5–7 years, one in three new public buildings could have a photovoltaic energy plant installed.

3. The extension of basic research into energy technologies, from nuclear fusion to solar energy, is imperative in order to ensure a continuous flow of new knowledge. This is clearly a task where government action is required. A reduction in the funds made available for such research (6. EU Framework Program for research, technological development and demonstration [2002–2006]) is definitely not the right approach.

4. Beyond that, governments should engage in the areas of appropriate education and infrastructural creation of competence for new energy technologies.

5. The total energy consumption in residential households and in connection with mobility is still rising. There are a number of approaches available to influence the present negligence of consumers concerning energy consumption. The instruments of regulating households as main origin of energy consumption, emissions and waste are far less developed than in the industrial domain (e.g. the IPPC directive stipulating the use of the "best available technology").

Consumers are not only restrained by the lemming effect but also by information deficiencies preventing them from making sound decisions. Therefore, we recommend effective and credible labeling including the "greenpricing" of electricity from regenerative sources. The poor success of previous approaches can be explained by the overflooding of the market with labels, leading to the failure even of official labels to clarify sufficiently the difference between sustainable and non-sustainable product. In some cases, this was a result of industry lobbying, in others it was caused by a lack of practical differentiation criteria.

6. More promising could be to bank on social and organizational innovation in order to accelerate the diffusion of energy innovations in residential households. An example of such innovation is the experiment of supply companies – often partnered by public institutions and enterprises – to position themselves as service providers. Such a step frees them from the pressure to sell more and more energy. Instead, they can offer profitable services for the efficient use of energy.

7. In the transport area, more energy-efficient providers have to build more comfortable and faster logistics or mobility chains, compared to private and commercial motor vehicles. Higher market shares cannot be achieved by improving the individual components, but only by revising the entire transport process. This requires innovative packages, for instance the link-up of rail traffic with car sharing and information services.

8. Still, in this case too, only looking at technology development is not enough. At least equally important is the compatibility with existing infrastructures and processes as well as the integration with the electricity network, both on the local level (for compensating discontinuous resources such as wind) and in the European arena (for instance by channeling hydropower produced in Scandinavia to the South during winter and using photovoltaic energy from Italy in the North, during the summer months).

   For fuel cells, overcoming the "chicken and egg" problem (no vehicles without hydrogen distribution, no hydrogen market without vehicles) is a crucial precondition for the success of this promising technology. Its true advantages, however, with regard to CO_2 intensity will only emerge when hydrogen will be produced from regenerative energy sources.

9. Energy issues must regain the highest political priority.

10. Beyond that, we call for the foundation of an "alliance for sustainable energy innovations" organized as a network.

Political enforceability of sustainable energy innovations

This study is not limited to the development of options for action; it also examines their political enforceability. We explicitly look at the interests of different social groups of agents (politicians, consumers, companies, environmental organizations etc.), where we will touch on three aspects in particular: the aims level, the choice of instruments and the strategy for enforcement.

Sustainable innovation in the energy sector lead, as a whole, to the long-term growth of social wealth; at the same time, however, it requires transformations in the social aims system as well as reforms that appear less attractive, at least in the short term, to some social groups (including essential parts of the energy industry). Given the interests of the groups of agents, we cannot expect the spontaneous emergence of a broad coalition for action concerning sustainable innovation in the energy sector. Each such group (enterprises, consumers etc.), on the other hand, disposes of a certain scope for action towards sustainable innovation, which they can realize without having to give up their principal interests. If we succeeded in exploiting this scope consistently, we could create a dynamics of reform that could take us beyond the status quo (which is not conducive to sustainability). The question remains, how to initialize and organize this process.

Concerning the choice of instruments, one has to examine the enforceability of the instrument mix invigorating sustainable innovation in the energy sector at the lowest possible social costs. In the political process, economically efficient instruments stimulating innovation, such as certificate solutions or eco-taxes for environmental protection, have turned out be less than attractive.

The instrument mix proposed here has a better chance of enforceability. It calls upon subsidy solutions, self-regulation and information instruments (labels etc.), because these do not imply immediate, perceivable strains on well-organized groups of agents. Hence, in the short term, sustainable innovation policies must make use of the potential of these instruments in particular.

Innovation-guided policy will have to face resistance, too, but not as much as allocative policies based on taxes, regulation or certificates. The promotion of sustainability policies requires a clearly defined process for formulating binding targets and for ensuring maximum engagement and the creation of capacities towards a platform, on which groups (even if they support opposite positions) can learn (to improve) co-operation, exchange experiences, report on their learning successes and become an integral part of a global network with a shared vision of the future.

New institutions are usually treated with skepticism. In contrast to specialized authorities, from central banks to antitrust regulators, sustainability affects every aspect of life; it cannot be separated from the core of democratic politics. Nevertheless, our analysis has shown that it often suffices to link-up existing institutions in a network, in order to integrate the concept of sustainable development into their particular competences.

We therefore propose the foundation of an "alliance for sustainable energy innovations", which should focus on three objectives:

− Increasing the public awareness of the divergence between energy demand and growth limitations and the potential of sustainable energy innovations for „pushing out the boundaries" if this potential is exploited more urgently,
− identifying obstacles (e.g. inefficiencies) to an accelerated realization of sustainable energy innovations and the promotion of new solutions, and
− building a database and a center for the transfer of knowledge on sustainable energy innovations, for an easy access to every information on specific energy innovations, partners for co-operations, consultancies etc.

The members of such an alliance could be:

− Enterprises and industry federations (e.g. producers of solar and "conventional" energy, energy customers),
− scientific research institutions,
− energy agencies and institutions for technology transfer, and
− consulting and service companies with innovative, creative ideas.

The more members the network includes, the more valuable will it be for the individual member. Ultimately, the alliance could cultivate contacts beyond Europe, especially with developing countries, where the real struggle for a sustainable energy system is fought.

Responsibility for the "energy hunger" of the developing countries – How can sustainable energy innovations help here?

The measures proposed here may be judged on a global scale, but the measures themselves largely focus on the EU. However, we would not meet the criterion of intergenerational justice if we failed to examine to what extent sustainable energy innovations could level recognizable „north-south slopes".

The principal features [characterizing the situation in the southern hemisphere] concern the dearth of competence and infrastructure, the limited commercial supply of energy and the inefficient use of energy sources, especially fire wood. Many modern technologies for energy production from regenerative sources – from wind power, biomass and solar radiation, in particular – ought to be employed in those countries. However, before such technologies can gain practical relevance, competences and infrastructures have to be built. In the sparsely populated rural areas outside the towns, decentralized technologies are often much more useful than centralized provision. In this field, too, far-reaching, targeted measures are necessary.

A multitude of international organizations – primarily the World Bank and the Global Environmental Facility (GEF) – endeavor to support sustainable energy systems and innovations in developing countries. The EU, on the other hand, suffers a particularly low profile in this area: Energy does not play a significant role, not even an institutional one. This has to change, and there have already been various good proposals (e.g. the G8 Task Force, whose well thought-out suggested were, unfortunately, rejected).

Apart from that, the number of successful examples show that global enterprises play a much more active role in technology sharing, for instance through direct investments (e.g. production plants for wind turbines). Obviously such enterprises would rather be guided by policies than venture on unknown territory independently. Therefore, the EU would have to make the effort of integrating energy issues and energy technologies with her development policies more closely than in the past. In this way, a framework and incentives for more investment in energy innovations and their development by the private sector would emerge.

However, development aid, technology transfer etc. will only become effective if the industrialized countries themselves manage successfully to transform their own energy systems. Hence, energy innovations in industrialized countries are a precondition for sustainable energy systems in developing countries. This realization takes us back to the starting point of our analysis: Even if the global dimension of the energy question is indisputable, most energy-relevant decisions are taken on national, communal or even individual levels. Therefore, to promote sustainable energy innovations, we need a long-term, international engagement by all agents on all those levels, from enterprises to nongovernmental organizations, from scientists to national administrations.

1 Problem Definition, Tasks, Procedure and Derivation of Recommendations for Action

1.1
Problem definition

The energy debate is full of paradoxes. On the one hand, global energy prognoses and scenarios show a stubborn "upward" trend – a continuous growth of energy consumption, caused by growing economies and populations, appears inevitable, with only the scope and the rate of growth still controversial. The usual pattern in this is stagnation in Europe and rapid growth in the emerging countries (and the US). On the other hand, a further growth in energy consumption increasingly faces limits: The obligation to reduce the "greenhouse gas" CO_2, which is almost exclusively caused by energy consumption, is the most prominent example, but by no means the only one. Also, the shift of oil production (back) to the politically unstable Middle East, the volatility of oil prices, the immense investments for the development and production of non-renewable energy sources, which also thwart the development of renewable energy sources, and the limited assimilative capacities of planet earth, e.g. as a receptive medium for pollutants from energy-intensive production, are important arguments. Both in science and in international declarations on principles, the problem of intra- and intergenerational justice (within populations living now and in relation to future generations) arises incessantly, while there appears to be no reason to believe this aspect to be of any particular influence on energy-political decisions taken by governments, businesses or even consumers. Only about 10–15% of the German population have ever heard of the term "sustainable development", a concept aiming to bring together economic, ecological and social criteria for decisions in all areas, thus bringing issues of justice within and between "north and south" and between present and future generations into the discussion.

Looking for ways to deal with these problems in a meaningful manner, one sees the "principle of hope" at work, mostly the hope of "innovations" opening up new sources of energy and supposedly revolutionizing the efficiency of energy usage. The concept, or rather, the hope associated with innovation becomes the "dummy variable" between the various predictions and discernible restrictions.

1.2
Tasks of the working group

It was against this background that the interdisciplinary working group established by the european academy faced the task of systematically examining the question: what is the true potential of innovations in the energy sector and is the sustainable development we strive for really achievable through them? In regional terms, we proceed, foremost, from the situation in the countries of the European Union, without neglecting the wider perspective and the repercussions especially for the developing countries.

The question demanded a clear focus on the intersection of the three central issues: Energy, innovation and sustainable development. On each of these subjects, there exist whole libraries of all kinds of studies; the number of possible "side tracks" ahead of the group was virtually boundless. While the importance of energy consumption for all areas of life may be undisputed in principle, it is impossible to investigate every facet and distant effect. This applies, in a similar way, to the subject of "innovation", which has been a long-running theme in the discussion of economic theory from as early as Schumpeter (1911). The concept of sustainability, on the other hand, has become a fashionable, political term only recently, though it is already the subject of countless controversies. Everybody lays claim to it, or tries to exclude others from it. Hence, our intention was not to embark on a comprehensive treatment of the three subject fields; neither could many of the debates in this context be pursued beyond the point where they ceased to be instrumental in developing the strategy proposed by this study.

Among the first tasks was, therefore, the establishment of theoretical connections between energy, innovations and sustainable development, based on a precise formation of terms and concepts, and thereby to define the exact focus of the investigation from there. Differentiation was called for, to make the investigation manageable. The concept of sustainability, for instance, involves not only ecology and economy, but also a social dimension, which is addressed here by dealing with issues of employment, aspects of development politics and questions concerning the security of procurement[1].

Eventually, however, the social dimension touches every area of personal and communal life: The individualization of society leads to a further increase in the number of single-occupant households and hence in the energy consumption per capita. The public promotion of owner-occupied home building as well as policies concerning road building and traffic have an effect on energy consumption. Every attempt to make such long-term effects of energy innovations part of the scope of this study, to analyze and assess them, would be inviting failure not only due to limited resources and time restrictions, but also because many developments are simply unforeseeable – a point we will support with arguments at a later stage. The social dimension of sustainability represents a boundary condition for the options for action dealt with in this study, in the sense that any energy scenario that puts into question basic elements of our political, economic, social and cultural development models appears unacceptable. However, this does not exclude the question of individual as well as institutional responsibility for the further development of this wealth-generating model under the various restrictions in a world that is more crowded, with a human population of ca. 10 bil. by the year 2050, and an immense backlog in the third world countries.

---

[1] "Security of procurement" ('Beschaffungssicherheit') should replace the term "security of supply" ('Versorgungssicherheit'), which marked a certain position in the energy discussion of the past.

The "Archimedean point" of most expert reports is the question: what are the ends that should be pursued? From there, recommendations are developed for how to achieve these ends. The working group followed another route: It examined the international and EU legal standards, from which the postulate of sustainability and, accordingly, the course to be set for energy policies can be derived. It is a matter of political standards set by institutions that are legitimized to do so, not of "science-immanent" definitions. It is difficult, though, to derive operational ends of action from such standards, which are not specific in terms of quantities. Hence, out of the numerous energy scenarios prepared by various organizations, a middling scenario cited by the World Energy Council (WEC) and the IIASA (International Institute for Applied Systems Analysis) was chosen as a reference point or benchmark, which was then combined with a "moderate scenario" from the IPCC (Intergovernmental Panel on Climate Change). This scenario was compared to an estimation of the potential of innovations in the areas of energy efficiency and renewable energy sources, with all the uncertainties immanent to such calculations. Nevertheless, one would have to resort to a number of extreme assumptions in order to *avoid* the statement that, with the present "business as usual" approach, the sustainability aimed for is missed by a wide margin. Since giving policy recommendations in an advisory function was part of the remit of the working group, we could not stop at merely stating a need for action.

1.3
Deriving recommendations for action

The political debate about the environment and the economy certainly does not suffer from a shortage of discussions and investigations concerning the instruments available for energy policy implementation. Still, abstract assessments of instruments are of little help, since they will always look "optimal" under the given assumptions. In the view of the working group, instruments can be assessed only if the objective *and* the context are clearly defined (which is often difficult enough, because objectives are formulated vaguely, and the context changes frequently). Even under the most favorable conditions for our considerations – having a quantitative reference point for the sustainable consumption of energy – the high degree of complexity[2] and the uncertainties in estimating the effects of instruments often lead to a situation where reliable, quantitative assessments of innovation potentials in the energy sector become impossible – especially in view of the, inevitably, distant time horizon of our study (to 2050). For innovations, this "uncertainty" is constitutive as far that they are characterized, in principle, by their open-endedness with regard to results and processes: We do not know what we will know.

In the group's opinion, the lack of quantitative estimates in the recommendations for action is a disadvantage however only to some degree. For the crucial point is whether one succeeds in inducing an accelerated learning and development process in the energy sector through economic and scientific activities on the European level, taking possibilities for innovation beyond the borders of nation states.

---

[2] On a precautionary note, it should be pointed out again that complexity is not just the fashionable word for complicatedness. Complexity is defined as the multitude of possible states of a system.

Progress so far has been rather sluggish, because of the large, long-term employment of capital required, but also due to mental and institutional barriers especially where energy has been provided through a regulated monopoly.

However, changes in areas with such far-reaching effects as those of the energy sector always face conflicts of goals, too. Certainly, there are "win-win" solutions, through which different criteria improve simultaneously, for instance if an innovation reduces both the costs and the environmental strain of energy provision and, at the same time, does not show any negative societal effects. More likely, though, are the cases where conflicts between different goals emerge. Hence it is important to acknowledge such conflicts – and to search for instruments and combinations of instruments that reduce them to the extent that they become politically feasible (for instance by forming a majority coalition of actors in favor of a practical compromise).

These are the criteria on which our examination of previous experience and knowledge was based. Especially in the light of current research, we arrived at results, in some cases, that are clearly at variance with the prevalent discussion – e.g. concerning the role of subsidies or voluntary agreements at different phases of the innovation cycle in the energy sector and for specific segments of energy innovation (see chapters 6 and 7).

Still, a strategy must set priorities, however necessary it may be not to lose sight of the differentiation and diversity of energy consumption. It seemed a reasonable decision, therefore, to assume strategic priorities where both the energy consumption is particularly high and the potential for improvement is greatest. This brings energy-intensive processes in industry and households into focus, as well as energy-intensive products and – especially in view of the great potential of the EU in general and Germany's responsibility as the second largest export nation, in particular – the accelerated development of regenerative energy sources. The central question was always: How can existing potentials for innovation be transformed into the widely used "state of the art" technology more quickly?

One could think of other priorities, another focus or a different structure of the recommendations (e.g. determined by the classic fields of politics). In this respect, our recommendations may (and should) well meet with opposition. Still, we believe that through a cooperation of elites, this strategy will also allow the formation of a majority coalition of actors. For giving recommendations for action was only half of our work. The other half was examining the chances of their implementation.

1.4
Structure of the study

Firstly, a sound investigation requires a clearly defined terminological and conceptional framework. The second chapter provides such a framework for the three central concepts of sustainability, energy and innovation. In this, our intention is not to add a few to the, perhaps, more than 200 different definitions of sustainability already existing. As everybody knows, definitions cannot be right or wrong; they can only be adequate or inadequate. For the clarity of our analysis, it is important to distinguish between *sustainability* as a general guiding idea, which initiates a searching and learning process with a practical intention, and the more concrete concept of *sustainable development*. The latter takes into account particular specifications and thereby serves, principally, in this study as an action-guiding concept for identifying the potentials and possibilities for action pointing to a path that might take us closer to the ideal of sustainability.

Next, the different concepts, from "weak" to "strong" sustainability, are analyzed. The working group decided to use the concept of "critical sustainability" as the reference criterion, since it is most suitable with regard to the long-term strategy of innovation-based sustainable development. On the one hand, it avoids any dangerous watering down of the idea of sustainability such as the mutual substitutability of any forms of capital, as assumed for (very) weak sustainability; on the other hand, it takes account of the fact that concentrating on only a few, but crucial and, in this sense, critical economic "crash barriers" or "bottlenecks" can yield realistic strategies for sustainable development.

The second area to be specified comprises the term energy. From the perspective of science, the role of energy as a non-substitutable key resource for humanity goes back to the fundamental importance of the energy flow within the biosphere. Consequently, the colloquial terms "energy demand" and "energy consumption" describe the human demand for energy of a high quality ("exergy") and its degradation into energy of a lower quality, respectively.

In the area of innovation, we go even a step further. Apart from the terminological specification, we briefly report the status of the scientific discussion about this subject, in order to provide a basis – especially for those readers who are not familiar with the economic discussion – for the later chapters, which are characterized rather by economic issues and where we turn to results of the scientific discussion, most importantly in the recommendations for action.

The third chapter deals with basic normative criteria for consideration and decision-making, which, far too often, are introduced into the discussion only implicitly. Consequently, depending on the underlying premises, one can arrive at very different interpretations with regard to the shaping of sustainability. The result is the, at times, confusing debate that can be observed in the present situation. Crucial for anyone's analysis is the uncertainty of the knowledge base in the areas relevant to the sustainability discussion. However, since environment-related goals (land use, climate change) as well as goals relating to the energy system itself (reliability, openness of options) and the availability of resources (security of procurement) must be considered, a broad set of goals can be formulated, on which decisions are to be based.

In the fourth chapter, we examine the deficits and points of reference for a sustainable energy system in the context previously specified through critical sustainability. Our starting point was the body of standards stipulated by international and constitutional law and relating to sustainability, especially climate protection and procurement security. These shall serve, in a concretized form, as reference points for evaluating the global energy system.

According to all the relevant prognoses, sustainability is missed, at least on a global scale, not just in terms of critical sustainability but with regard to every concept of sustainability one finds reported. The (global) trend towards deregulation rather aggravates the sustainability issues (which is not an objection against a stronger orientation towards competition in this formerly monopolistic sector but only points to the accompanying measures necessary). From the supposed "back-stop technologies" – energy from nuclear fusion or fission, respectively – decisive contributions to solving the problem cannot be expected within the period to 2050.

Two models of operationalization are developed: The "time of safe practice", as a criterion for the time available before a new (sustainable) route of development must be achieved, and the reference point of a global mean continuous power consumption of 2,000 watt per person (2,000-Watt benchmark) – also compatible with the moderate IPCC scenario S450 for the sustainable provision of energy, which is based on the intelligent use of energy and on regenerative energy sources, without expecting the citizen to sacrifice any comfort.

The technical potentials, which can be exploited for reshaping the energy system, are discussed in chapter 5. Above all, apart from efficiency potentials, the potential of regenerative energy sources deserves mentioning here. These potentials are summarized in some development perspectives that would allow to achieve the formulated goals.

In chapter 6 we analyze the conflicts of goals that impede a sustainable provision of energy. Here, the key issues are concerns about employment and competitiveness, unwelcome effects of deregulation and the promotion of innovations, undesirable consequences for development policies and, finally, limits to intervention set by the 'sovereign' consumer. Taking account of such conflicts of goals is no less important than knowing the normative rules of balancing stipulated (wisely) by EU legislation.

Chapter 7, finally, focuses on the strategy recommendations: How can we close the gap between the "trend", or "business as usual" approach, and the exemplary reference standard of a global energy system in agreement with an average power consumption of 2000 Watt per person?

Having specified precisely the strategic objectives, we focus on four problem constellations: The accelerated market introduction of sustainable energy innovations, the faster diffusion of best available technologies through voluntary agreements, the promotion of not yet marketable regenerative energy sources through a balanced set of measures, and activating the consumer by stimulating a "demand pull" to promote the chances of sustainable energy innovations in the market. Since our mandate was not only to present recommendations for action but also to analyze the chances of their implementation and develop suggestions for increasing the chances of their realization, we propose, in chapter 8, an "Alliance for Sustainable Energy Innovations".

While the focus of the previous chapters was on the situation in the countries of the EU, in chapter 9 we examine the importance of the industrial countries for the developing nations that are now "catching up". After all, every effort on the European level would be of little use if the large, and growing, majority of deprived people were to copy our current habits of energy consumption. In this area, also, the exclusive reference to the government would fall short of addressing the problem properly, for businesses too can and should assume an active role in technology sharing.

2 Terminological and Conceptional Foundations

2.1
Sustainability and sustainable development

2.1.1
Terminological differentiation

Particularly since the early 1980s, the terms "sustainability" and "sustainable development" have been associated with hopes expressed in the discussion about global conditions for a life consistent with human dignity. Hence, these terms carry decidedly positive connotations. Following its original emergence in late medieval forestry in Central Europe, the concept of sustainability was extended from a circumscribed subject area of the lasting cultivation and utilization of forests – first from the perspective of timber economy, exclusively, but since the 19th century increasingly, also in view of the wide-ranging ecological functions of forests (water balance, local and regional climate, species diversity, soil preservation, recreation etc.) – to more and more other spheres.

If the extension of the sustainability concept from the regenerative resource, timber, to other renewable resources, such as fish stocks, still seems justifiable through systematic reasoning – and it present only few problems in content –, the publication of the report of the World Commission on Environment and Development (WCED), "Our Common Future" (1987) fundamentally changed the situation. This report extended the criteria of sustainability and sustainable development[1] to a world economy and world community, which are interconnected in many ways and characterized, above all, by the fact that they rely only to a small extent on regenerable resources that can be used in a truly sustainable fashion. Instead, exhaustible resources such as coal, mineral oil and natural gas are predominant. These resources cannot be used in a sustainable manner *per definitionem*, at least in terms of a concept of inventory, since each unit of an energy carrier consumed in a certain place today cannot be used again to the same extent at another place or another time (*temporal rivalry in consumption*). From that perspective, the two central normative cornerstones of sustainable development in the conception of the World Commission on Environment and Development (Brundtland Commission), i.e. *intragenerational justice* between the countries of the north and those of the south, and *inter-*

---

[1] For the German version of this report, we translated this term as „*Dauerhafte Entwicklung*". Later, however, the term „*Nachhaltige Entwicklung*" gained acceptance, not least because the English term "sustained yield" or "sustainable yield" is itself a translation of the German *"Nachhaltiger Ertrag"* (Nutzinger/Radke (1995a), p. 16).

generational justice between people living today and future generations, attain special importance.

However, the chances of realizing sustainable development improve as soon as one, reasonably, replaces the static concept of sustainability, which focuses on inventories of resources, by a *dynamic* concept of sustainability, which is not oriented along (mostly given) stocks of resources but along the possibilities of overusing those stocks. The concept of inventory and the concept of use are "inescapably" coupled only in a static economy devoid of any technical progress. If one further takes into account the phenomenon referred to as *innovation*, i.e. the successful implementation of new solutions in the production and marketing of goods and services (see chapter 2.3 for more on this subject), the seemingly absurd extension of the concept of sustainability to a global economy relying predominantly on exhaustible energy resources does not look like an utterly hopeless venture anymore: If appropriate innovations succeed in reducing the use of exhaustible resources per unit of power in production and consumption, so that the use-up rate of such limited stocks can be lower in the future, then the possibilities of using these resources can be maintained and sometimes even enhanced despite the diminishing inventory of resources.

Yet of course one has to consider that sustainability is a complex and challenging concept, which is not open to simple technical solutions. For instance, we cannot overcome the scarcity of exhaustible resources simply by transforming the entire planet into a plantation for regenerative raw materials. Still, even if innovations cannot augment, or just keep constant, the physical stocks of exhaustible resources, innovations may well increase the value of existing inventories for present and future generations. Thus, the otherwise impenetrable conflict arising from the rivalry in the consumption of such resources becomes "solvable", at least in principle, in the sense that there can be meaningful possibilities of use, on which all concerned parties can achieve consensus. Nevertheless, considering the current trends in the areas of energy consumption, environmental degradation, private consumption and population development, the possibility of such consensual ways of using resources does not mean that we can really find a course of sustainable development that meets the many competing demands to an equal extent. This study, therefore, attempts to determine the chances of such a course of development in a reasonably realistic manner. In doing this, we cannot and must not leave out necessary changes in long-term trends.

At first, however, we are going to draw the distinction between sustainability as a general regulative idea, and sustainable development as a possible practical application of this idea. This distinction aims at a pragmatic determination of possibilities for action, not at a collection of recipes for concrete measures, starting from the existing boundary conditions and the foreseeable developments (see chapters 7 and 8 for details).

2.1.2
Sustainability and sustainable development

The scientific discussion around five experts' reports on "basic regulatory issues of a sustainable policy" (Gerken 1996) prepared for the German Federal Ministry of Economics shows in an exemplay way how difficult it is to concretize the concept of sustainability without either specifying it in its contents to such a degree that the necessary "future open choice" is affected, or leaving it so imprecise in its substance that the concept is rendered useless even as a reasonable heuristic for the necessary search for sustainable courses of development and degenerates into a mere catchword.

In the following, the concept of sustainability is interpreted as a general guiding concept, which initiates and accompanies, with a practical intention, a search and learning process, whereas the more concrete concept of *sustainable development* is regarded, in principle, as a guide for action. This is not supposed to be a definitive or even generally binding definition of sustainable development; it is rather intended to determine a course that takes us as far as possible towards the idea of sustainability.

1. Sustainability as a guiding concept – in common with other guiding concepts such as freedom or justice – comprises both a descriptive and an explicit, a priori normative component. In its more general form, though, it is more appropriate for pointing out forms of managing the economy that work against sustainability; it is less suitable for identifying measures that would promote sustainability. However, if "sustainable" or "lasting" development is the issue, as it was for the World Commission on Environment and Development (WCED 1987), more concrete stipulations need to be drafted. Hence, a development guided by the concept of sustainability must comprise more than a general heuristic. As a matter of fact, the "Brundtland Commission" itself was eager to point out that, taking into account technical progress, which is of analog importance here as are innovations concerning products and processes, as well as aspects of justice between the countries of the north and those of the south, as well as between present and future generations, it views such lines of development as realistic possibilities. Understandably, the Commission did not embark on detailed descriptions in content and, instead, confined itself to referring repeatedly to the equally justified claims of all human beings – wherever and whenever they may live – to a life in accordance with human dignity.

2. Inversely, when warning about "political preconditions" that are too concrete, the risks involved with an unlimited variety of interpretations of the term "sustainability" must not be underestimated either: On the one hand, even an intellectual search process requires sufficiently clear ideas to guide the search. On the other, there is the serious danger that the conceptual vagueness of the fundamental guiding concept conceals necessary conflicts and balancing acts and feigns substantial agreement where conflicting interests have still to be settled.[2] This does not do justice to the actual problem situation. Even if sustainability as a guiding concept can never be defined conclusively – the same is true for the concept of health called upon by Homann (1996, p. 38) in his critique of the concept of sustainability –, the intuitions associated with these concepts by no means exclude certain specifications, especially negative ones; they may even require them in some cases. The guiding concept of sustainability does in fact also serve

---

[2] This last risk affects parts of the report of the World Commission on Environment and Development (WCED 1987).

to exclude, already on the heuristic level, certain developments, processes and measures in obvious contradiction to this concept, which initiate or discernibly favor policies resulting in the lasting destruction of nature.[3] Still, further concretizations, e.g. the setting of certain targets, must be added, if specific measures with regard to a desired sustainable development are to be derived from the general concept of sustainability.

3. The current discussion on climate shows signs of avoiding such hazards by defining certain maximum boundary conditions or "crash barriers". This becomes particularly clear in the assumption on which the IPCC scenarios are based and according to which the rise in the mean global temperature should be limited to 0.1° C per decade.[4] Against this, one could argue that planet earth could also "handle" more rapid climate changes. However, even if one accepts (provisionally) this precondition of limiting the speed of climate change, there is a further need for discussion – which has not been met consensually so far – about the consequences of greenhouse-relevant emissions for global mean temperatures. Further research and discussion are required here, but there we face the problem that we may know with certainty "only at the end of a decades-long process of searching, learning and experiencing" that, in a condition of incomplete information, we have brought about a situation where irreversible, counter-sustainable climate changes have occurred. Thus, the guiding concept of sustainability demands not only a process of social discussion, but also the readiness, in a situation of insufficient knowledge, to take appropriate action (e.g. a reduction of greenhouse gases) to prevent situations that might *ex post* prove to work against sustainability – if it can be shown through balancing processes that the prevention strategies are proportional (see chapter 3).

4. Nevertheless, the fact remains that sustainability, as a guiding concept, is better suited to exclude actions working against and endangering sustainability than to produce policy recommendations that are sufficiently concrete. Therefore it is not enough to regard sustainability as a mere guiding concept. It is reasonable to distinguish between *sustainability* and *sustainable development* in two stages by first defining the meaning of the general guiding concept before further inquiring which identifiable trends (most notably in the provision and use of energy, and in the innovations sector) will ensure an obvious failure to deliver this guiding concept. The two-stage nature of this formation of concepts meets the requirement of openness for a general guiding concept of sustainability and allows for the definition in content of a workable sustainable development, which further characterizes a course of development guided by the idea of sustainability.

5. In this sense, sustainable development, in contrast to sustainability, is an explicitly *constitutive* concept. If we take seriously the objective of sustainable development as a concept guiding our actions, we have to agree on the course to be set and the orientations required so that we do not fall short of the guiding concept

---

[3] The importance of always keeping in mind the dual character of sustainability as a descriptive and, at the same time, normative characterization is obvious here.

[4] This assumption of the IPCC is based on the available knowledge about the speed of natural, geological climate changes.

of sustainability. From this, very specific recommendations for action can result, especially for the provision and use of energy and for the process of innovation, which are guided primarily by recognizable bottlenecks for future developments.

2.1.3
Different concepts of sustainability

Considering the multitude of efforts to define "sustainable development" – by now, there are more than 200 of them after fifteen years of scientific and political discussions –, one cannot but admit that this concept is still very vague or, at times, even mired in confusion.[5] In the present discussion of the problems surrounding sustainability, a first approach leads to the observation of three different ways of dealing with the varied meanings of "sustainable development": Apart from *sheer disapproval* (because of the supposed "wooliness" of the concept) and an *all-embracing strategy* (burdening the concept with everything that happens to suit one's purposes), there is another possible attitude, which is shared by the working group: the effort to deal with the concept in a *productive* manner and to define it as precisely as possible according to scientific criteria. This involves comparing various possible definitions of the concept and asking: What are the concrete conclusions in each case, with regard to the research question central to our study? This path is taken in neoclassical environmental economics and, above all, by ecological economics, the "science of sustainability" (Costanza 1991). One has to find a balance between overdetermination and underdetermination of this concept; one should neither burden it with specific requirements which meet the most stringent ecological criteria, but make it an unachievable ideal, nor should one leave it so vague that it can mean everything and effect nothing: In principle, the concept must be operational.

In the following, we shall introduce some productive definitions of this concept of sustainability, and discuss them briefly. Whichever definition of *sustainable development* one considers, they all share, basically, a normative orientation towards a principle (however defined in detail) of intergenerational justice: A development is sustainable, in the words of the Brundtland Commission (WCED 1987, p. 46), if it meets the needs of the present without compromising the ability of future generations to meet their own needs. This is, typically, the central quote used in the "saturated" world, whereas the very next sentence in the report tends to be forgotten:

> It contains within it two key concepts: the concept of "needs", in particular the essential needs of the world's poor, to which overriding priority should be given; and the idea of limitations imposed by the state of technology and social organization on the environment's ability to meet present and future needs.

The principle of intragenerational fairness, as the second normative element of sustainable development, is closely related. This connection is inevitable, both from the ethical-philosophical perspective and from the practical point of view. From the ethical perspective, one could say that human beings, who feel responsible for the

---

[5] A survey of the spectrum of definitions can be found in Enquête Commission (1998) and Kopfmüller et al. 2001.

well-being of their descendants, should feel at least equally responsible for their contemporaries' well-being (see e.g. Daly and Cobb 1989).

The discussion within economics – in neoclassical resource economics as well as in ecological economics – focuses on issues of intergenerational distribution. For a long time, the economy of resources reduced the problem to a mere optimization calculus without properly taking into account the problem of justice. It is thanks to ecological economy that the questions of justice, which are implicit for instance in the routine discounting of future benefits and costs, have been pointed out explicitly. As we cannot deal with the various positions concerning intergenerational justice in any detail in this report (see Turner, Doktor und Adger 1994, p. 267, on this subject), we only note briefly that the acknowledgment of obligations of the present generations to those in the future represents a normative decision on principle, for which there is no binding, definitive reason, even if there are many plausible arguments in its favor (see chapter 3).

Many economists have tried to define the meaning of sustainability in terms of intergenerational justice by the requirement that a constant capital stock has to be passed on to later generations, where this capital stock consists, in first differentiation, of man-made, material capital, on the one hand, and "natural capital", on the other. This not only raises further problems of definition, such as the question as to what exactly natural capital is supposed to be and how it is supposed to be measured,[6] but also leads, foremost, to the controversial question of how those different types of capital relate to each other. Are there any substitutive or complementary relationships between material capital and natural capital? Starting from this very question, one distinguishes today between at least five concepts, described by the terms very weak sustainability, weak or critical sustainability, and strong and very strong sustainability:

Very weak sustainability is discussed in at least two different forms. The weakest form only requires that the annual gross national product (GNP) – i.e. the valued, periodic utilization rate of a not necessarily constant, aggregated capital – should not decrease over time. We could call this form "extremely weak sustainability". The second form, sometimes referred to as "very weak" in the literature, sometimes as "weak sustainability", stipulates that the value of the total aggregate capital shall be constant over time, assuming a perfect mutual substitutability between material and natural capitals. Pearce und Atkinson (1993) proposed an indicator for measuring this (very) weak sustainability, according to which a country is on a (very weak or weak) sustainable development path if the savings are larger than the total depreciation of both material and natural capital. Empirical studies show that numerous countries do not even meet this criterion of very weak sustainability (Pearce & Atkinson 1993, Atkinson et al. 1997). Ultimately, this (very) weak sustainability goes back to neoclassical resource economics, to the model of Robert Solow (1974), in particular, and its extension by John Hartwick (1977). Justified doubts about the "sustainability" of such a weak indicator have been voiced by, among others, Faucheux et al. (1997).

---

[6] In contrast to material capital, the measurement and valuation of natural (including human) capital is still in its beginnings. The World Bank, among others, is increasingly engaged in collecting appropriate data (see World Bank 1995, 1996).

The advocates of a *weak sustainability*, or rather a *critical sustainability* or *quasi-sustainability* (for instance the "London School" around David Pearce, or Nutzinger/Radke 1995b) argue that (very) weak sustainability overlooks the fact that natural and material capitals are not completely but only partly substitutable. Hence they introduce the concept of a *critical natural capital*, which marks limits of substitutability and takes account of the fact that the elements of the natural capital are not only input for the economic process but, to a certain degree, prerequisite for human life and hence for economic activity as such. In this view, certain "keystone species" or "keystone processes" are indispensable for human survival and, therefore, cannot be replaced by man-made material capital. Hence, this concept of weak sustainability (i.e. *critical sustainability*) demands that limits, "safe minimum standards" or a precautionary principle are set, by which any allowable economic consideration must be confined (see chapter 3).

Safe minimum standards, like "crash barriers" (WBGU 1996), load capacities and critical stocks (Endres/Radke 1998), so prominent in the discussion about sustainable development, are in fact environmental standards (Streffer et al. 2000). Environmental standards are *set* in view of a specific purpose; they are not just *determined* through scientific research. However, results of scientific research always play a role in the process of setting environmental standards. When studying for instance the possibility that the Gulf Stream might "stop flowing", one can use a simulation model to determine under which conditions this could occur. Assuming the preservation of the Gulf Stream is desired, the model computation (which must be subject to debate itself, with regard to its method) presents the basis for formulating values that *shall* serve as crash barriers limiting the relevant actions. In this example, preserving the Gulf Stream is the environmental quality goal, which serves as guidance for setting environmental standards.

Notwithstanding this setting of limits, the concept of *critical sustainability* allows for degradations of natural capital *exceeding* the safe minimum standard, as long as these are balanced by the growth of other forms of capital. To this extent, critical and weak sustainability are indeed overlapping. From the perspective of *strong sustainability*, on the other hand, such balancing would not be tolerated. Due to the high uncertainty of the available knowledge concerning the intricacies of ecological systems, about the irreversibility of interventions in ecosystems and the adequate assessment of natural capital, which is not entirely possible, this capital should remain constant on a scale of *physical indicators*. In practice, the demarcation between weak (or, as we call it, critical) and strong sustainability is difficult, as Turner et al. (1994) have pointed out, for the requirement of a "constant" natural capital may also be derivable from the requirement that "safe minimum standards" are indeed safe.

Finally, the concept of *very strong sustainability* calls for the limitation of the total *scale* of economic activity as part of the ecological system. Here the *throughput* of matter and energy is required to be minimized, not least in view of thermodynamic implications, most notably the law of entropy (see chapter 2.2). Still, it appears hardly possible to translate such abstract physical rules of minimization into concrete suggestions for action.

A more practical approach, in the view of the working group, would be to extract from the scientific analysis concrete "safe minimum standards", which would have to be observed in the future development of economies. Indirectly, one of the goals of strong sustainability, i.e. limiting the anthropogenic increase in entropy, would be pursued, too, by adopting *critical sustainability*.[7]

All types of sustainability presented here, from extremely weak to very strong sustainability, imply a certain form of intergenerational fairness, namely the bequeathing of a certain natural potential to the following generations. They differ in what this inventory is composed of, basically with regard to the underlying assumptions about substitutability or complementarity of man-made and natural capital in each case: Is it merely a constant gross national product (GNP) (*extremely weak sustainability*), is it an overall constant but randomly combined inventory of natural and material capital (*weak sustainability*), is it a minimum of ("critical") natural capital (*critical sustainability*), or is it a constant natural capital (*strong sustainability*) which must be maintained and passed on to the next generation?

Having such an abundant choice of definitions, why do we call upon a concept of *critical* sustainability for the following discussion? The answer is: Considering our goal, which is a long-term strategy of innovation-supported sustainable development, the concept of critical sustainability is most appropriate because, on the one hand, it avoids dangerous dilutions of the idea of sustainability – such as, e.g. for (very) weak sustainability, the random, mutual substitutability of different forms of capital –, and, on the other hand, it takes into account that realistic strategies for sustainable developments can only be achieved by focusing on a few, but decisive and in this sence critical "crash barriers" or "bottlenecks" which have to be observed in economic activity. Hence, we are neither looking for paths of development that are realistic but discernibly countersustainable, nor do we seek paths that would fulfill the highest ecological aspirations while discernibly lacking any chance of prevailing. Furthermore, the link between the two concepts, sustainability and innovation, clearly points to the central idea of this study, according to which humans, through their capacity for innovation, can indeed replace natural capital by material capital, within the critical boundaries mentioned above – an assumption that the history of mankind has confirmed in many ways.

2.2
Sustainability and energy

In our view, another fundamental concept apart from sustainability plays a central role: the concept of energy. Why did we choose energy, in particular, as a crucial subject area for sustainability? In reality, our society depends on a large number of other non-renewable resources presently not used in a sustainable way, at least from a long-term perspective. These include all natural minerals, but also water, air, soil and the diversity of species inhabiting the biosphere. Beyond that, immaterial assets, especially the knowledge gathered in the course of millennia, and the organizational forms of living in societies or, in a word, human culture is an important factor for sustainability.

---

[7]   Other efforts of concretization discussed in the context of sustainability are the so-called management or utilization rules (going back to Daly; also see Nutzinger/Radke 1995b) and the ecological set of ends, **E**(lements of the biosphere)-**S**(elfregulation potential)-**H**(omeostasis), see Hampicke 1992, p. 314-322.

Confining ourselves to material goods, we conclude that in many respects there are no limits to human inventiveness; but there are boundary conditions, which must be regarded as an unavoidable given for innovations too. In the following, we will briefly illustrate that energy – the energy flow, more precisely – is one of these basic conditions. This finding justifies the choice of energy as a central subject area for our project. A society does not survive if it neglects the creation of a sustainable energy base, as in the history of humankind only such cultures have survived which succeeded in organizing their energy supply in a sustainable way.

2.2.1
The laws of thermodynamics and the concept of energy

Energy may dominate our everyday life and constitute an important production factor in economic theory; from the physics point of view, however, it is a rather abstract entity, which can only be defined accurately in terms of a differentiated mathematical model. Historically, the concept of energy was initially defined simply as the "potential to perform work". In that sense, of course, energy is not conserved; this is why the notion of "energy *consumption*" has become common usage.

The connection between the, at first, entirely different concepts of "energy" and "heat" was clarified only in the 19th century, with the formulation of the first law of thermodynamics stating that energy is preserved, i.e. it is neither created nor destroyed, but just transformed from one form into another. (At the beginning of the 20th century, the concept of energy was extended by Einstein in his theory of special relativity, which includes mass as a form of energy.) Hence, energy consumption actually means energy *degradation* i.e. transforming valuable or available energy (exergy) into lower-value or non-available energy (anergy). The boundary between exergy and anergy is not absolute, but depends on the system considered. For instance, water at a temperature of 20 degrees Celsius in an environment at zero degrees contains usable energy (exergy), while this would not be the case at an ambient temperature of 20 degrees. The common physical units of energy and energy flows are defined in box 2.1.

BOX 2.1: Energy and power: joules and watts

The *amount* of energy is measured in **joules (J)** or **kilowatt-hours (kWh)**. The energy flow [or flux] *per unit of time* is called power, with the unit 1 **watt (W)**. 1 watt is defined as **1 joule per second**. Kilowatt-hours per day or per year are also in use as units of power. The same units are used for all forms of energy (fuels, electricity etc.).

2000 watts continuous power (during 24 hours per day) is equivalent to:
- 2000 joules per second, or
- 48 kilowatt-hours per day, or
- 17,500 kilowatt-hours per year, or
- 1,700 liters of fuel oil or gasoline per year.

The biosphere and, hence, human existence is governed not by the first but by the second law of thermodynamics, also referred to as the *entropy law*, which is about the convertibility of different forms of energy: Energy of a certain quality can be converted completely into an energy form of lower quality. The reverse process, though, is only possible in part, i.e. with a finite efficiency. The maximum thermodynamic rate of conversion is the *Carnot efficiency*. This maximum efficiency explains for example limits on the production of electricity (high-quality energy) from heat (lower-quality energy) in thermal power plants, or the physical boundary conditions for the operation of heat pumps. If the second law did not exclude it, the total energy requirements of humankind could be met e.g. by reducing the temperature of the oceans very slightly.

2.2.2
Energy systems in the biosphere and anthroposphere

Living creatures are structures of a high molecular complexity or order. As a result of the entropy law, any life needs a continuous flow of degradable energy to maintain this order, i.e. to defend against chaos. The biosphere draws almost all its energy from solar radiation. As the energy flow from the sun is not available uninterruptedly, life would be impossible without some mechanism for storing energy. The process of *photosynthesis* converts solar radiation energy into storable chemical energy. The average global rate of photosynthesis in the biosphere is equivalent to a power of 130 TW (1 TW = 1 terawatt = 10^{12} watt). This is only about 0.1 per mil of the total solar energy flux reaching the earth's surface (box 2.2).

The physiological energy demand of a human being amounts to ca. 10 million joules per day and person, corresponding to an average power intake of ca. 100 watts per person or, globally, a power of 0.6 TW. Since the transition from any trophic level to the next higher one involves losses of typically 90%, humankind

---

BOX 2.2: Global solar and anthropogenic energy flows

	Per surface area (Watts per m^2)	Total (Terawatts)
Total solar radiation on earth's surface	240	122,000
Global commercial energy use	0.02	12
Physiological energy required by humankind (100 Watts per Person)		0.6
Biological primary production		
Total (land + sea)	0.25	130
Land (reference area excl. Antarctica)	0.44	65

would require ten times the above power (6 TWs) from global primary production. In reality, the share is even larger, because humans take in part of their food in the form of meat, i.e. from an even higher trophic level.

Beyond his physiological energy demand, man has invented, in the course of history, various methods to set aside part of the huge solar energy flux for other human needs. The development of agriculture represents the most significant pre-industrial innovation. By clearing woodlands, agriculture increases primary production artificially, and raises the share of biological production useful for humans by selecting and cultivating suitable plants. The cultivation, harvesting and utilization of agricultural products required a great deal of mechanical energy (work), which came from man himself, from working animals, or from the use of natural forces (wind and water power). Aside from the mechanical energy required, thermal energy was always needed, too, for preparing food as well as for heating. The demand for heat was met almost exclusively by burning biomass (wood, animal waste, etc.); fossil fuels did not play a role, although in certain places coal had been known for a long time.

The term *energy system* (of a country or earth as a whole) designates the entire structure of the primary energy resources in use and of the infrastructure for their distribution and conversion into final energy, and of the specific demand structure of the so-called energy services (see box 2.3). With regard to the quality of the

---

BOX 2.3: Energy conversion chain and gray energy (also see chapter 5, fig. 5.1)

Energy for human use passes through a conversion chain, which consists of the following links: Primary energy carrier – primary energy – secondary energy – final energy – useful energy – energy service. To illustrate the links of this chain, we have chosen the example: "running an electric locomotive with power from a coal-fired power plant":

Primary energy carrier	Coal as a raw material in the ground
Primary energy	Mined coal
Secondary energy	Hot steam in a thermal power plant
Final energy	Electricity
Useful energy	Force on the drive wheel of the locomotive
Energy service	Transport of people and goods

A country such as Germany imports energy not only as oil, natural gas etc., but also in the form of goods that have been produced using energy. This energy is referred to as **gray energy**. Conversely, gray energy contained in goods is also exported. Countries with a large service sector and not much heavy industry show a trade deficit in gray energy, which the national energy statistics must take into account. For Switzerland, for instance, the trade deficit in gray energy is estimated to represent 25% of the direct consumption of primary energy.

energy form, the distinction between the demand for heat or work, respectively, plays a special role, as well as the differentiation between stationary and mobile demand and the function of electricity. Together the supply and demand structures determine the potential for changing an existing energy system.

While the *preindustrial energy system* was based almost exclusively on the use of solar, i.e. renewable and in most cases locally available energy resources, the situation changed fundamentally with industrialization, in two respects: firstly, through the invention of the steam engine and, secondly, through the discovery and utilization of electricity. The steam engine, and other thermic engines that arrived later, allowed for the first time the conversion of thermal energy into work (even if with considerable inefficiencies, in accordance with the second law of thermodynamics). With this, the demand for thermal energy resources grew to such an extent that the utilization of fossil fuels (coal, later mineral oil and finally natural gas) became, and still is, the most important pillar of the *industrial energy system* (see chapter 4). Energy production from nuclear fission has not changed this situation, despite its nearly fifty-year history; its contribution to global energy has still not reached 3%.

In contrast to fossil fuels, the introduction of electricity into the energy system did not add a new source of *primary energy* (see box 3). Electricity is a form of *final energy*. Its great significance lies in the fact that it can be produced from various primary energy sources (fossil and nuclear fuels, water, wind, photovoltaic energy and others), it can be transported easily over vast distances and can be used, as an energy form of high quality (exergy), to meet any demand for thermal or mechanical energy. Electricity's mean global share in energy supply amounts to 20% today, and the trend points upward. Electric power is mostly (60%) produced in thermal power plants fired by fossil fuels (predominantly coal). These power plants are responsible for nearly 30% of the global carbon dioxide emissions into the atmosphere.

2.3
Innovation and sustainability

2.3.1
Basic context

Notwithstanding their very different meanings, the concepts of sustainability and innovation have important characteristics in common, which are equally important for our subject matter as are the crucial differences between the two concepts. Both terms, "sustainability" and "innovation", represent concepts that are often poorly defined but still endowed with positive connotations in most cases. They are often used as actual or supposed "problem solvers" in crises where a possible, immediate course of action is not discernable.

Thus, "sustainability" – most notably in the triad of ecological, economic and social sustainability – often serves as the "magic word" that seems to show a way out of the actual or, more important, impending crises caused by human economic activities that over-utilize natural capital both on the input side (natural resources) and on the output side (assimilation capacity of natural systems). Something similar can be seen in the use of the concept of innovation, in particular if it is widened beyond the realms of technology and economy, to encompass society as a whole, and becomes "social innovation"; for innovations are characterized by exactly this: They overcome or at least relax previous restrictions by introducing new products and processes.

Not every innovation, however, makes a positive contribution to sustainable development. At the outset, innovation is just a source of structural change, social and economic, contributing to economic growth and fluctuations in economic activity. To what extent innovation as a whole promotes employment and/or reduces environmental degradation is not predetermined from the start but depends on the social framework, price ratios or the choice of technology: In principle, it can be shaped.

Various theoretical and empirical studies suggest that innovation, on the whole, offers a positive contribution to sustainable development:

- The new theory of growth puts the emphasis on *increasing returns to scale*, i.e. a general growth of the factor productivity (and thus the environmental and resources' productivity).
- The structural change driven by innovation, among other factors, leads to a reduction on environmental strains ("gratis effects").

In the following, we will briefly outline the aspects of innovation research relevant to this study.

2.3.2
The concept and types of innovation

The question whether innovation processes can be influenced by policies is, of course, important with regard to our recommendations for action. It can be answered in a meaningful way only if reliable knowledge about the determining factors (innovation determinants) as well as the process phases and the effects of various innovation activities are available. This is illustrated in fig. 2.1.

In the following sections, we briefly report the status of the discussion on process/product innovations, including the specific issues concerning the energy sector. Then we will examine / investigate how their success in the market place (market entry, diffusion) could be supported, i.e. which policy alternatives would be effective, which obstacles would have to be removed, or in other words, which institutional innovations are necessary (see also chapters 7 and 8).

The concept of innovation in economic theory was crucially influenced by the works of Joseph Schumpeter (1911, 1942). In this tradition, innovation is defined as the implementation of new problem solutions or factor combinations in the market. The decisive criterion is commercial success, which is identifiable, however, only ex post. Innovation must be distinguished from *invention*, which is a new finding, technological idea or solution that can usually be protected by patents or other intellectual property rights. Not every invention can prevail in the market and thus form the basis for an innovation. An invention is neither a necessary nor a sufficient precondition for innovation, although invention activity clearly has a positive effect on innovation, since the frequency of innovations rises with the frequency of inven-

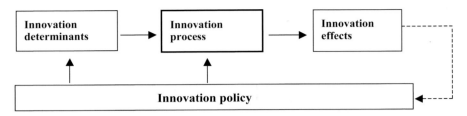

Innovations determinants	Innovation process	Innovation effects		
Taxes	R&D (Research and Development)	Economic sustainability	Social sustainability	Ecological sustainability
Regulation	Invention	Growth	Distribution	Material/energy flows
Education system	Prototype	Employment	Social safety	Land-use patterns, diversity of species
Competition policy	Market introduction	Price stability		
Infrastructure		External trade balance		

Fig. 2.1 Innovation process, innovation determinants, innovation effects

tions. Only the *diffusion*, i.e. the broad distribution of an innovation opens up its economic potential. Diffusion is driven not only by the innovator, but also by *imitation* which is often accompanied by further improvements.

Based on the criterium "significance", two types of innovation can be differentiated: *incremental innovation* (improvement) and *breakthrough* (fundamental innovation). Further differentiation is possible (see e.g. Kemp 2000, where three classes of innovation are defined, "incremental innovation", "radical innovation" and "system innovation"), but each new category brings additional problems of delimitation. "Everyday" innovation activity is dominated by incremental innovations, which have no sweeping effect in the short term but lead continuously and over a longer period of time to far-reaching changes (the car is a good example here). Breakthroughs occur only after long, irregular intervals, but then they effect fundamental upheavals in economy and society.

Referring to the subject of the innovation activity, normally two types of innovation are defined:

— **Process innovation:** new production plants/methods/processes, new forms of organization, opening-up of new markets and supply sources.
— **Product innovation:** new functions, qualities, forms/designs of goods and services.

More recently, the concept of innovation was extended to other aspects of economic and social development beyond this traditional use in economic theory. In summary, two other types of innovation can be identified:

- **Institutional innovation:** changed framework conditions (autonomous central bank, regulation regime, national plans for sustainability etc.), prevailing especially in the international competition of economic and social systems.
- **Sociocultural innovation:** changed values, lifestyles, consuming patterns, (working) time patterns, needs, preferences etc. in a society.

All these types of innovation are relevant in the context of sustainability. Similar to this definition, "environmental innovations" are defined as "those techno-economic, institutional and/or social innovations (…) that lead to a qualitative improvement of the environment (…) whether or not these innovations would be advantageous from other, namely economic perspectives, too" (Klemmer et al. 1999, p. 29). All types of innovation usually discussed in the context of environmental protection are included: "end-of-pipe" technologies, process-integrated environmental protection, product-integrated environmental protection and function orientation (Kurz 1996).

2.3.3
The innovation process: Inside the Black Box

1. Phases of the innovation process: In the simplest case, the overall economic innovation process can be subdivided into three phases:

More differentiated approaches distinguish between a larger number of phases (including R&D, prototyping etc.), also taking feedbacks into account. The relevance of different models of the innovation process, with respect to policies in support of innovation, is that they reveal interfaces, and thus transfer problems and possible bottlenecks or barriers, which can be starting points for such support or for innovation-supporting reforms of the political framework. With his pioneering work, "Inside the Black Box", Rosenberg (1982) initiated the economic analysis of innovation processes (also see Freeman/Soete 1997, Dodgson/Rothwell 1996). Innovation research has produced numerous findings about the determining factors (determinants) of the innovation process. Nevertheless, there remain considerable gaps in the knowledge especially about the relative importance of the different determinants. In the following, we briefly outline the results that are of particular interest for our project, especially those that help to identify obstacles to innovations (in the energy system).

2. Innovation through a "technology push": Innovation can be "produced" systematically. The most important input factor is expenditure for R&D (manpower, equipment, materials etc.). The higher this input, the higher the output in the form of innovations – albeit only with a certain probability, for innovation involves risk and uncertainty. In a simplifying manner, one can distinguish between basic research and applied research. Basic research is not directly oriented to current problems, although it is clearly influenced by them. The perspectives of future economic exploitation are always considered when main research initiatives are endorsed. In this sense, decisions about central projects of basic science are important in setting the course at the early stages of the innovation process. They constitute a certain self-amplifying dynamic in development (e.g. in the utilization of nuclear energy or in biotechnology,), a specific "scientific-industrial complex". Applied research attempts to apply known principles to new questions. The direction of applied research in enterprises is strongly influenced by economic conditions and expected changes in these conditions, most notably concerning the relative prices of the labor and energy factors. As early as the 1930s, the hypothesis was formulated that changes in the factor price ratios alter the direction of the innovation process (Hicks 1932, Popp 2002). When labor costs, for example, rise consistently faster than capital costs, this induces labor-saving technological progress (rationalization). Innovation efforts are increasingly aimed at the factor showing the (relatively) greatest increase in cost. Accordingly, rising energy prices should induce more innovation activity towards energy savings. This expectation has been confirmed convincingly in the study by Popp (2002), who arrived at the result, for the US, that rising energy prices – even if such rises are determined by environmental taxes or regulation – "encourage the development of new technologies that make pollution control less costly in the long run" (Popp 2002, p. 26).

3. Innovation through "demand pull": The innovation process is driven, ultimately, by the prospect of (exceptionally high) profits. With free (unrestricted) competition, these profits always are only head-start profits, meaning they will be destroyed by imitators. Such profits can be achieved when existing needs are met in a better way, or when new needs are raised successfully. Hence, innovators react to (previously undiscovered) demands of (potential) customers and – as dynamic entrepreneurs – they also create new demands. Politics can have a crucial influence on the kind and extent of the demand. In the extreme case, government can prohibit new products/processes (or at least delay their licensing) and thus prevent the development of an innovation. The government can support an innovation by

– defining the new problem solution (e.g. as "state of the art") – and thereby accelerating especially its diffusion;
– making existing solutions more expensive (e.g. through taxes on fossil fuels) or even prohibiting them (e.g. toxic, persistent chemicals);
– acting as a pioneering buyer (lead customer), who is prepared to pay a high price initially (although the product quality may still be patchy), thus creating the conditions for mass production and learning effects.

4. Cost advantages in the innovation process: The innovator creates a new market and gains the chance to shape it substantially (e.g. through setting standards). Especially where high fixed costs are involved, every potential competitor will find it difficult to achieve access to the market (first-mover advantage). The innovator is also the first to profit from learning effects. When companies introduce a new technology, high changeover and adaptation costs are incurred initially. These costs fall to the extent to which the company and its workforce learn and implement minor improvements (learning curve; also see the excursus in chapter 7.8.3). Experience, skills and tacit knowledge are accumulated, which also provide protection against quick and simple "copying" (imitation). Falling learning costs and an expanding market volume allow the unit costs to decrease (self-amplifying process). Hence, governments frequently try to support the build-up of "national champions" to compete internationally and make their first-mover advantage benefit the entire national economy (growth and employment effects).

Learning curve effects yield a self-amplifying process – larger market volume → falling costs → lower price → larger market volume, etc. –, which strengthens the market position of the innovator, as long as falling unit costs are always passed on through

---

Box 2.4: Learning curves

Learning curves capture the phenomenon, which is well proven empirically (e.g. in the aircraft industry and in electronic chip and computer production), that unit costs decrease with the total number of units produced:[8] From everyday experience in the production process, firms learn how production costs can be reduced (*learning by doing* (Arrow 1962)). Obviously, such potentials cannot be exploited by research alone. The production process itself brings about both minor (incremental) and significant improvements (process innovations).

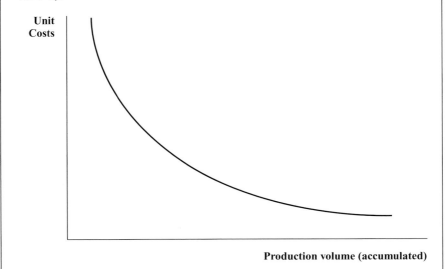

---

[8] Learning curves are usually displayed in a diagram with the total volume produced (sold) assigned to the abscissa, on a logarithmic scale, and the price per unit to the (likewise logarithmic) ordinate. From that representation, a linear relationship between the (accumulated) volumes and the cost can be derived.

the unit price. This penetration strategy makes it more difficult for potential competitors to access the market. However, this strategy may involve low profits and the innovator will finally cancel it. This increases the chances for newcomers to enter the market, especially if there are "spillover effects" (e.g. through the enticing away of qualified, experienced workers). Otherwise, a permanent monopolization of the market could result. The learning curve does not take into account product improvements emerging from user feedback. However, the increasing quality of a product is at least equally important as a factor for accelerating its diffusion.

Among the results of empirical studies on learning curve effects in the energy sector are the following (OECD/IEA 2000b, Williams 1994):

- Doubling the (worldwide) sales of photovoltaic modules leads to the price falling by 18 %.
- For (Danish) wind turbines, doubling of sales leads to the price of wind energy decreasing by (only) 4 %. Still, the learning effects are markedly greater (18 % EU, 32 % USA) if one considers not the individual plant but the entire process of producing electricity from wind energy (including the selection of locations, maintenance etc.).
- The learning rate for the production of electricity from biomass amounts to 15 % per doubling of the production volume.

Government measures supporting market introductions, and thereby facilitating the onset of the learning curve dynamics, appear to be well justified because

- they have positive external effects (know-how spillover to other sectors, reduced environmental strains);
- they compensate for "first-mover advantages", e.g. market barriers, enjoyed by old technologies and development paths, which were established in some cases through public subsidies and regulations.

Government can contribute to the learning costs (i.e. it can instigate "learning investments") by stimulating the demand of private pioneers (subsidies, regulation) or by presenting itself as a buyer (state procurement purchases). Such policies are aimed at the innovator, in the first instance, but they must also never fail to consider potential newcomers, because otherwise they encourage a lasting monopolization. Supporting policies must be discontinued as soon as a technology is mature and the learning effects achievable (by incentives) become minuscule.

5. *"Embodied" and "disembodied" technical change:* The diffusion of new problem solutions takes place through new products and facilities that incorporate the new "state of the art" technology (embodied technical change). Major gross investments are required to drive the rapid "modernization" of the capital stock. Therefore, conditions favoring investment must be created (e.g. through investment grants or favorable depreciation rules). In this way, the replacement of obsolete facilities (power plants, vehicles etc.) can have positive environmental effects. Shortening the service life of old facilities is, however, not a general strategy for sustainability, because

- new (e.g. automotive) products are not always moreenvironment-friendly;
- Reduction of pollution occurring in the utilization phase must be compared to additional pollution during the production and future disposal of facilities.

A decision on whether a "long-term product" is indeed superior – or what its optimum service life is, economically and ecologically – can only be made on the basis of the overall balance (*life cycle assessment*, LCA). Hence, it must be examined in each individual case whether accelerating the diffusion through investment incentives really assists a sustainable development. Apart from the "embodied" type of technical change, there is also "disembodied" technical change, which is not linked to physical capital and which concerns, above all, the management and economic organization of the production process (organizational innovation). This includes management techniques, forms of distribution, logistics concepts, innovations in capital markets (e.g. venture capital), car sharing, energy contracting and establishing markets for recycled materials. Such organizational innovations enable efficiency gains that do not require any significant additional expense on resources (for production and disposal). There are no obvious, concrete ways of promoting this type of innovation, but (government) regulations will most likely play a role in this area.

6. *Trajectory-dependence of the innovation process:* Once the innovation process has moved into a certain direction and proceeds on a technology path (trajectory), any change, like for instance the switch from direct to alternating current, can only be effected at high costs. Along such a trajectory, incremental, continuous improvements take place, which contribute to its further consolidation. Once the trajectory is fixed, changes in market conditions, especially in price ratios, can effect only marginal adjustments, which do not immediately lead to a change of trajectories. In some cases, a change of the technological course only succeeds after a long time, when a new window of opportunity opens. In a "revolutionary" situation (bifurcation point), when the old technology path is left and (two or more) new paths open up, it is crucial which technology is the first to gain a critical mass of advocates/customers. In the case of network effects, (almost) all customers will eventually switch to this technology. At this point, government support for market introduction can develop a considerable effect, either positive or negative. Network effects are at work if the benefit of a product or a service to each consumer rises with the number of consumers. A single telephone is useless; a small telephone network with a few participants is of modest benefit; only if many people own a telephone, does the technology display its full potential. Network effects also occur with technologies in which the net is not immediately recognizable, e.g. with the use of a widespread PC operating system (which ensures compatibility, service and the availability of application software). Decisions on infrastructures and standards agreed by private or public sector institutions (DIN, ISO etc.), as well as regulations, play an important role in the emergence of a (new) technology path. By removing compatibility problems, standardization can contribute to a faster diffusion, though it can also lead to a premature end of the competition between different technical solutions – before the superior variant is clearly discernible. Standards set by the private sector must be judged critically, not least because they emerge from a process that is dominated, as a rule, by large, established companies or associations. Therefore, we are dealing not just with the competition-neutral solution of technical problems, but with organizations asserting their interests: Whoever determines and dominates the standard will have a competitive advantage. It is not always the more efficient technology that prevails (and thereby determines the

course of innovation for years to come). The most commonly quoted examples of suboptimal solutions that won through and became established (see Fleischmann 2001) include the video cassette recording standard VHS, developed by JVC, over Betamax, by Sony, and the QWERTY typewriter keyboard over the Dvorak keyboard. Innovation policies introduced by governments can actively set the technological course (e.g. nuclear energy, air & space, biotechnology) – and thereby promote expensive mistakes.

7. *Regional clusters:* Regional concentrations (clusters) of resources and competences, which often emerge as accidents of history, enhance the macroeconomic innovation activity. Such clusters are mainly the result of

- geographic proximity, leading to lower communication and mobility costs;
- social proximity, favoring informal contacts on many levels and the development of trust.

Innovation research tries to explore the conditions for the emergence of clusters, for instance through analyzing successful regions such as Baden-Württemberg, in Germany, or the Po basin in northern Italy. Based on such analysis, innovation policy can attempt to create new clusters. Universities, start-up centers, or the (subsidized) location decision of a pioneering enterprise can serve as the "crystallization point" of a cluster. If (sustainable) innovation policy relies on cluster formation (regions of competence), increasing regional disparities (differentials between centers and the periphery) can lead to effects which are not compatible with sustainable development (e.g. increasing mobility).

8. *Cooperation* (R&D joint ventures, strategic alliances): The success of innovation processes depends crucially on the cooperation of various agents. This cooperation can be organized in more or less formal arrangements . Cooperations offer important advantages, compared to mere market or hierarchical coordination. Collaborations extend across industrial sectors and can be of regional up to global dimensions. With increasing division of labor, innovation increasingly comes from suppliers, who are given general performance specifications instead of simply a design drawing. They are part of the learning network, which also encourages collaborations of subcontractors (system contracting). In some sectors, the innovations introduced by plant manufacturers (e.g. power plants) are of crucial importance. Ideally, such networks combine the specific advantages of large corporations, on the one hand, and of small and medium-sized enterprises (SME), on the other, in the innovation process. However, issues also arise concerning competition policy, such as buyers' power and specialization cartels. Again, governments can initiate or invigorate cooperation (e.g. through joint research projects).

9. *International links in the innovation process:* Due to globalization, innovation becomes ever more dependent on worldwide information concerning successful problem solutions found in other parts of the world, and on the international transfer of such solutions for application in a multitude of markets. Intensive foreign-trade relations produce learning effects, which in turn enhance innovative competence. Different elements of the innovation process are characterized by different degrees of international mobility. Basic knowledge diffuses very quickly, the diffusion of product innovations takes longer, and that of process innovations is even slower; some skills (precision, reliability, engagement) may not be transferable at

all. Because of the rapid diffusion of basic knowledge, it becomes increasingly difficult for each investor (public or private) to appropriate completely the returns of his investments in education and research (the "appropriability problem"). Concerning developing countries, there have been intensive discussions in the past about the extent to which these countries need special, "adapted technologies", for instance because they do not have the qualified workforce for running and maintaining complex facilities, or because western technologies would trigger negative employment effects. This debate appears to be dissolving due to the fact that the developing and emerging countries demand the latest technologies. Hence, the aim must be to ensure that the efficiency potential of such facilities is realized under conditions of continuous operation (e.g. through the support of [workforce] qualification measures).

10. National innovation systems: Since the 1980s there have been efforts to describe the factors determining the extent, direction and speed of an innovation in a countryin their context and to refer to them by the term "national innovation system" (NIS). It describes, in a systematic manner, the network of relationships between agents and institutions involved in the production, transfer and utilization of knowledge (see e.g. OECD 1999). Businesses and their ability to innovate are the central element of any NIS. Apart from that, the kind of relationships between the enterprises and between businesses and other institutions (especially government) are important. This approach, when applied to the comparison between countries, brings national features into the open as relative strengths and weaknesses. The strengths of the German innovation system are in the area of higher-value technologies rather than in top technologies (see BMBF 2000 and ZEW et al. 2000). The international leadership of the German environmental (technology) industry, where mostly higher, but not necessarily top technologies are applied, fits into this general picture. It is characteristic of the German, as well as the Dutch and the Danish NIS that the high population density combined with high levels of prosperity have led to high expectations from the populations concerning the environment. Furthermore, strong trade unions and other political agents have their demands concerning the social compatibility of innovations. After the creation of the Single European Market, though, the importance of the NIS has diminished considerably. While regional characteristics (clusters) remain in place, the European framework is increasingly replacing national patterns (also see appendix 3).

2.3.4
Factors determining innovation activity

The extent, direction and speed of innovation activity depend on the macroeconomic and social framework conditions as well as on the reactions and strategies of businesses (see Nill/Petschow/Jahnke 2001). We could list almost any number of such framework conditions. The difficulty is to identify those that crucially influence the innovation process as a whole and its individual phases. On this, economists have arrived at a far-reaching consensus, which is however, not always supported by unambiguous empirical evidence (see Kurz/Graf/Zarth 1989, Becher et al. 1990). The totality of the framework conditions (institutional structures) in a country at a given time – which together characterize the extent and type of the innovation activity – is referred to as the "national innovation system" (NIS). By studying the differences and similarities of various NIS one can identify success factors and obstacles and derive policy recommendations. Some important aspects have already been mentioned in the previous sections. Beyond that, the relevant literature emphasizes the following points:

- Intellectual property rights (appropriability problem): information, costs, infringements of patent protection (in the international context)
- Regulation: economicregulation (barriers to market entry in certain sectors e.g. transport, trades) and social regulation (safety standards, environmental protection, consumer protection, work safety)
- Taxation: income tax versus consumer taxes, losses carried forward and back to favor innovators with widely fluctuating incomes, generally low (marginal) income tax rates versus favorable depreciation conditions
- Financing/capital markets: bottlenecks in innovation financing (venture capital, seed capital)
- Innovation infrastructure: basic research, internet access, technology/start-up centers
- Education system/human capital (primary school to university, vocational training system)
- Transfer from university to businesses: cooperation, permeability, exchange of personnel
- Employees as a source of innovation: forms of participation, working time models
- Acceptance issues: technology assessment, forms of communication

Each of these keywords stands for a multitude of possible decisions concerning innovation policies. This will be briefly illustrated, using environmental protection as an example. The innovation effects of environmental protection depend primarily on its objectives and on the instruments chosen in order to realize them. Roughly simplified, the innovation effects of instruments of environmental policy can be summarized as follows (see in particular Klemmer 1999, Klemmer et al. 1999, and Zimmermann et al. 1998):

- Regulation ("command and control" approach) slow down the innovation process if a certain "state of the art" technology is rigidly codified (the "conspiracy of silence of the chief engineers"). An exception would be if ambitious standards, which cannot be realized with present technologies, could be credibly codified for a certain time in the future ("technology forcing", see Ashford 2000).
- As a rule, market instruments (liability, taxes, tradable permitts) favor innovations.
- With voluntary agreements (self-commitments), the innovation effect most often depends on the degree to which they are binding.

As a whole, however, the impact of the instrument mix employed in environmental policy on the macroeconomic innovation process should not be overstated. The "optimization" of the instrument set of environmental policy offers only a limited contribution to driving this process in a sustainable direction. A much wider approach is called for.

Empirical studies explicitly concerned with the determinants of "sustainable innovation" do not exist. Under the simplistic assumption that environmental innovations are also sustainable, one can fall back on the results of a ZEW study (ZEW et al. 2000):

- Regulations are of the greatest importance. The introduction of laws and the expectation of future legislation are the main impetus for environmental innovations. This supports the significance of the "technology forcing" approach. Still, one must note that the effects of such a regulatory approach on other fields of innovation – effects on which there exists only scant research – will more likely be negative.
- Cost savings (disposal, energy, materials and labor costs) are another important impetus for environmental innovations. Making energy and materials (and disposing of them) more expensive, e.g. through eco-taxes, should therefore have a noticeable effect on innovation activity. On the other hand, negative effects on innovation activity of a higher tax burden are another likely outcome.
- Another important point is the environmental awareness within corporations and in their surroundings (chances of image enhancement), in the sense that a more acute awareness of the environment can be of essential support for environmental innovation.
- The hope of opening up new markets plays a rather secondary role. "It appears to be the case that environmentally relevant product characteristics … are suitable for product differentiation, but not as essential product attributes *per se*" (ZEW et al. 2000, p. 83).
- The lack of reliabilty of the political framework concerning environmental policies, which is also among the reasons for the shortage of amortization possibilities, is probably the most significant obstacle to innovation. This highlights the importance of formulating binding sustainability targets (national sustainability strategy), which serve as an orientation guide and provide a solid basis on which expectations can be formed.

These results show that innovations reducing environmental degradation depend not only – and, it is suspected, not crucially – on financial support from government. Factors of regulatory policy (target definition, changes in price ratios, deregulation etc.) are at least equally important. This must be taken into account when formulating a strategy for sustainable innovation in the energy sector.

2.3.5
Sustainable innovation policy

In this section we present some general considerations on a strategy of sustainable innovation. First, we ought to clarify whether, and for what reasons, the state should intervene at all in a market economy to support innovation.

1. *Arguments for government support of innovation:* As a rule, private companies that are active in research and innovation do not manage to appropriate all the returns of their innovation activity.Beyond the profits arising for the companies, innovation also yields social returns (positive external effects, spillover effects). Hence, from a macroeconomic perspective, companies tend to underinvest in R&D. For this reason,

government intervention in favor of increased R&D or innovation activity enhances efficiency even in a market economy. Innovation support will most likely yield positive external effects if it focuses on basic research; the closer it is to the market (applied research, market introduction subsidies), it will lead to an increase in returns that can be internalized privately, whereas hardly any social returns (beyond the private ones) will remain. The closer innovation policy is to market introduction, the stronger is the selectivity of the support – and the problems with selecting the "right" projects ("picking the winners") and including SMEs (bias in favor of large corporations) become increasingly severe. Hence, there are generally good economic reasons for the state sponsorship of basic research, while in the case of applied research such support is justified only in individual, concrete instances. This position is put into context by the observation that microeconomic rationality can also lead to overinvestment in R&D ("patent race", parallel research etc.), and that government support would reinforce such inefficiency. Nonetheless, this objection should be of minor importance with regard to sustainable innovations since, by definition, the positive external effect will be considerable; apart from economic success, social and/or ecological improvements will be achieved ("double dividend", or even "triple dividend"). But even if one accepts the necessity of government support in principle, the following difficult "details" remain to be clarified: Which innovations (potentials, projects) shall be supported with which instrument, and to what extent?

2. *Innovation policy as a competition and regulation reform policy (accommodating policy):* If innovation is interpreted as the result of a social search and discovering process, to which a multitude of agents bring their knowledge and the results of which cannot be predicted in any detail, then innovation policy must focus primarily on "examining the institutional preconditions of innovative processes", meaning "essentially, the policy aims at identifying the restrictions on action that are suffered by the agents in the market and which obstruct, delay or limit a desirable (considering possible negative external effects) innovation process" (Wegner 2001, S 9 f.). At its core, innovation policy is competition policy aiming at the removal of a "dysfunctional regulation density" (restrictions of competition imposed by government) and anti-competitive practices followed by companies, as well as at strengthening the protection of property rights. In addition, innovation policy is to provide "complementary input factors" with public good characteristics. "Providing an efficient infrastructure" (Möschel 1994, p. 42) includes e.g. the education system and basic research. As far as the expenditure itself is well justified in terms of the allocation function of the state, government procurement policy can be used as a complement. "Institutional self-regulation" (Wegner 2001, p. 9) acts as an additional supporting factor: The market participants themselves create new institutions such as technical standards, quality and safety standards, and credit or sustainability ratings systems to assess corporations. In individual cases, government policy can intervene in the formation and development of these institutions, too, in a supporting or regulating form.

Innovation policy aiming at inducing or accelerating new technology paths can be justified and successful, especially if two constellations of conditions are in place (Wegner 2001, p. 21 ff.):

- It is conceivable that innovation policy is based on a transformation in individual preferences and that, for this reason, it only anticipates the consequences of such change (decline of traditional problem solutions). However, if this is the case, the difficulty lies in the problematic assumption that the state would have to recognize the transformation of preferences before the private enterprises notice them. This would mean that innovation policy is the attempt to "offset entrepreneurial competence deficits" (Wegner 2001, p. 11).

– Economic policy regulation imposes the abandonment of a technology path (decline of traditional problem solutions); the market agents increase their search efforts (e.g. their R&D expenditure) and are successful in finding a new (sustainable) technology path. The dynamics of innovation are not impeded but redirected, which incurs additional costs in the transition phase. This scenario depends crucially on the (unpredictable) innovation competence of the market agents. Here, too, preferences should be explored in order to suppress, as far as possible, any efforts of avoidance. "Economic policy would fall for a fatal illusion if it were to see the private competence for innovation as a regulatory resource to be exploited at will" (Wegner 2001, p. 22).

Hence, innovation policy should be based on as wide-ranging objectives as possible (e.g. increasing energy efficiency) and adjusted, largely, to the preferences of those to whom the benefits are directed (private individuals and companies) (e.g. with regard to the areas of need and the instruments available) (Wegner 2001).

3. *Phase-specific innovation policy:* Each innovation creates a new market with its typical life cycle (measured in turnover, unit sales etc.) (Heuss 1965):

I. Experimental phase: The product is developed until it is ready for the market, to which it is introduced butno unmistakable trend towards expansion is visible yet. The majority of new ideas and inventions do not proceed beyond this stage.

II. Expansion phase: The product prevails (exponential growth of market volume). The expansion of production volume is accompanied by falling unit costs, due to fixed-cost degression, learning curve effects and process innovations. Further product improvements and production efficiency gains are driven not least by the market entry of newcomers.

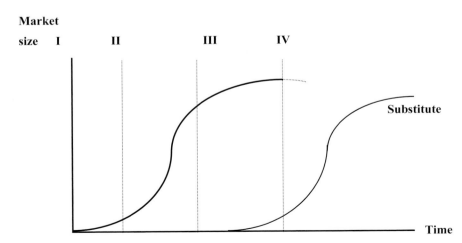

Fig. 2.2 Market development over the innovation process

III. Gestation phase: With growth rates declining (market saturation) and the innovation potential largely exhausted, the activity of enterprises shifts to product differentiation. Only product-oriented services (function orientation) offer chances for expansion. The entrepreneurs (as defined by Schumpeter) leave the market and open up a new market with another innovation.

IV. Stagnation/decline phase: Here we see the destructive effect of the innovation. Innovative substitutes destroy the market of the old champions, who are left to survive in niches. The necessary capacity adjustments are delayed, in many cases, by state (preservation) subsidies, which also hamper, indirectly, the success of the new champions.

Phase-specific innovation policy:

– At the onset of the expansion phase, government support for market introduction can accelerate the progress of the innovation – both through public grants for pilot and demonstration projects and through innovation-oriented procurement policies or subsidies for private first users. The question as to what instruments are most suitable is discussed in chapter 7. The curve in Fig. 2.2 would become steeper and peak (at the point of market penetration/saturation) further to the left. The peak can also reach a higher market volume – assuming domestic innovators succeed in achieving first-mover advantages (learning curve effects, setting the standard, etc.) and attracting international buyers.

– In the expansion phase, as well as in the gestation phase, the innovation becomes self-driven; (further) support from the state is not required. The difficulty is to recognize the right moment for the state to withdraw from its supporting role, and to carry out this withdrawal.

– In the phase of stagnation and decline, at the latest, the next innovation should commence its life cycle. To achieve this, enterprises must invest early (already during expansion and gestation) in R&D and prototypes. They do this for reasons of self-interest (using their revenues from the previous successful innovation, which now serves as a "cash cow"), though not always to an extent that would be desirable from the macroeconomic perspective. Government support can provide incentives, which help to ensure that the pressure in the project pipeline remains high, and that competence deficits do not become the limiting factor in the innovation process, which can be prevented especially through application-related school and education policies.

4. Dual strategy: A pragmatic approach to innovation policy combines the reform of the established R&D policy with a wider reform of the framework conditions.

– Innovation processes share regularities, and this enables us to derive starting points for economic policy measures. Support for (applied) R&Dand market introduction appears to be justifiable where competition has largely fulfilled its search and discovering function and the socially favorable, accelerated diffusion (reducing negative external effects and strengthening positive externalities) are left as the sole issue. This aspect of innovation policy is calculated to improve the chances of success (diffusion) of specific technologies (exploiting sustainable potentials). If there is an invention available, and if the assessment of its sustainability potentials (technology assessment) is positive, political instruments can aid in its development as an innovation, i.e. ensure its wide diffusion. It then remains to be clarified which instruments are most suitable for this purpose.

– General improvements of the framework conditions for sustainable innovation activities (e.g. regulation reform, tax reform, basic-research priorities) can help redirecting the innovation-seeking efforts of scientists and inventors. The knowledge and ideas pool within society (the stock of inventions) is enriched with ecology-guided contents and processes, the success of which is advanced by the state. This part of the dual strategy aims at a general increase in sustainable innovation activity, not at sector-specific potentials or certain types of innovation. And if this approach were to achieve a productivity growth in terms of material or surface area, this would be a contribution to a sustainable development, too. This element of the policy is relatively ineffective (in the short to medium term).

The two components complement each other. Insofar as the transitions between both types of innovation policy are fluid, any (at times ideological) "either/or" debate is unproductive in this context. The aim of sustainable innovation policy as a whole is to change the framework conditions in such a way that the chances of sustainable innovation potentials prevailing in the market are improved. The second component of the dual strategy for sustainable innovation is clearly hampered by the fact that successes only appear in the longer term, and they cannot be attributed directly to a specific change of the framework conditions. Hence, the political implemention especially of these types of reforms is a difficult endeavor (see chapter 8). The reasons for this are not only of a political/practical nature, however, but are also rooted in the conflicts of goals (see chapter 6) and the basic, normative decision problems discussed in the following chapter.

3 Normative Criteria for Evaluation and Decision-Making

3.1
Risk assessment and recommendations for action

3.1.1
Scientific policy consulting

In contrast to political lobbying, which is bound to certain interests, scientific policy consulting aspires to interest-neutrality (or at least to keeping a greater distance from organized interests, however legitimate these are in a pluralist democracy). The other difference is the transparency and systematic approach with which the results are dealt with and, most importantly, the recommendations are derived. These characteristics are called for especially in the case of interdisciplinary research, where the individual paradigms, assumptions or implicit normative pre-judgments cannot be identified in a (relatively) unequivocal manner as in the case of, for instance, economics research institutes with their known orientations, where a close correlation between the results of their research and their economic orientation is assumed.

The research group considers its work as part of the tradition of interdisciplinary technology assessment, which has proven its worth especially for environmental issues. Institutionalized scientific policy consulting concerning the environment has its origins in the scientific assessment of new technologies (United States Senate 1972, Haas 1975) and in the sociological treatment of the acceptance of large-scale technological projects (Renn 1984). These roots still characterize present efforts in this field, summarized under the label TA (for "technology assessment"; Bullinger 1994, von Westphalen 1997, Bröchler et al. 1999). Over time, the methodical foundations were extended by competences from (applied) ethics and the philosophy of science (Gethmann 1999, Grunwald 2000, Hanekamp 2001). Furthermore, institutionalized forms of TA today take into account the research desiderata of biology and medicine.

In the environmental sector, the very wide spectrum of scientific policy consulting is characterized by the challenge of integrating "subglobal" action with an assessment of global effects. Still, a certain confusion of aspects or options to be considered becomes obvious even in such limited contexts. Both developments point to the uncertainty of the knowledge base as, apparently, the central problem. For the energy sector, with which we are concerned here, both aspects are important with regard to sustainable development. The context for changes in the energy sector is a global one – the natural resources, for instance, have to be assessed on a global scale, in most cases, as well as the relevant emissions – but action is usually taken on a national or subnational level. The relevant changes in the environment are evaluated by means of models, whose predictive power can be assessed reliably only ex post – but recommendations are needed today. Conscious of this dilemma of environmental policy, the present study assembles the voices of science to enable well-informed decisions with regard to sustainable energy innovations.

3.1.2
Theoretical and practical perspectives

Action, in this context, is action under risk, meaning action that tolerates possible, unwelcome side effects. The possibility of such effects always refers to certain knowledge, a certain theory, or a certain model. Thus, the reliability of the relevant knowledge is decisive in risk assessment. However, the side effects looked at in the context of sustainable development are not just effects of a single action, but effects arising from an entanglement of actions and developments. This complication affects the problem of assessment, most notably in its ethical aspects, since it makes it difficult or even impossible to identify a correlation between the effects of action and any specific agent.

To assess the risks of climate change, one would have to study the individual elements of climate forecasts in theoretical terms, from the perspective of the philosophy of science. But not even such an investigation could be expected to remove the uncertainty of knowledge in that area: The uncertainty appears to be constitutive in this case.[1] The discipline that affords procedures for decisions under uncertainty is referred to as normative decision theory, which deals, in a general way, with comparisons between options for action in view of selecting such options (or the option) through which a certain end is achieved in the best possible way. If a particular state of the environment becomes reality, taking a certain option for action leads to a specific result.[2]

The various methods of decision theory can be classified according to what knowledge about the occurrence of certain environmental states is available. If the state of the environment is predictable, the decision is called a decision under *certainty*. If one can only cite probabilities for the incidence of particular environmental states, one deals with a decision under *risk*. Furthermore, if the possible environmental states are known, but not the probabilities with which they occur, the decision is one under *uncertainty*. Decision procedures for these three types of decision problems are available. For decisions under uncertainty, there is e.g. the risk-averse minimax rule (minimizing the maximum damage) or the risk-inclined maximax rule (maximizing the maximum benefit or, in this case, minimizing the minimum damage). However, these procedures require that the relevant environmental states, i.e. the range of conditions can be determined.

Hence, if the range of conditions cannot be formulated, these procedures are excluded. Robert Goodin (1982, p. 174) has described such situations (which apply

---

[1] See also the study of the European Academy: *Klimavorhersage und Klimavorsorge* (Schröder et al. 2002).

[2] In this way, a decision problem can be represented by specifying the options space, the state space and the results space.

e.g. to climate change) as cases of *profound uncertainty*, in contrast to the *moderate uncertainty* referred to above, where environmental conditions are actually known. The profound uncertainty of the underlying scientific problem is therefore an obstacle to the application of the risk-averse minimax rule, although it seems to be indicated, at first sight, because of this very uncertainty. Even the specification of a maximum damage – which would be excluded according to this rule – is in fact uncertain.

The epistemological difficulties surrounding the minimax rule do not affect the *moral* reasons initially tempting us to apply the rule e.g. in the case of climate change. These reasons are essentially the same as those with which John Rawls (1971/1998) supports his suggestion to use the minimax rule as a version of his well-known principles of justice, which are: (a) the moral asymmetry between damage and benefits; (b) especially if the individuals concerned will carry the long-term consequences of decisions under risk; (c) if these risks impinge on the vital interests of the individuals concerned; and, finally, (d) if those individuals did not make the relevant decisions or did not agree with them. When studying the risks of a climate change, we are dealing with a situation where criteria (b) to (d) are all relevant. Hence, we might not have the knowledge sufficient either to apply the minimax rule stringently or to compare costs and benefits precisely. This does not devalue the moral perspective, though, from which the application of this rule was suggested.

If we model the decision problem – should energy systems be rebuilt towards sustainability? – as a decision under uncertainty[3], with two options (sustainable restructuring or no sustainable restructuring of the energy system)[4] and two states of the environment (influencing climate change through restructuring or not), and if we further take into account the possible result that sustainable restructuring has no influence on climate change and, hence, does not prevent possible negative effects, then applying the minimax rule leads to the decision to abandon such a project. The two options are indifferent only if restructuring does not incur any costs.

On the other hand, even if restructuring involves positive costs, investigations concerning climate change are necessary only if the central objective of the (re)organization of the energy system is to prevent climate-related damage. The reverse argument reads as follows: By one-sidedly relating sustainable restructuring to climate change only, we burden ourselves with an obligation to give reasons that can hardly be met.

However, if such sustainable restructuring of the energy system is still desired, independent of the risk just mentioned, the costs of restructuring have to refer to all relevant criteria. Resource economy and energy economy may contribute further reasons for a sustainable reorganization. Considering the exhaustibility of fossil energy carriers, a reorientation towards regenerative sources of energy appears reasonable. With regard to the security of procurement, becoming independent of so-called crisis regions and taking precautions against the civil or even military risks of nuclear energy may be sensible aspects, too.

---

[3] Cf. Hanekamp 2003.
[4] In this way, the option space, too, is simplified to a straight yes/no alternative. Yet, it could also be widened, with a continuous "restructuring parameter" as the other extreme.

The reasons for sustainable restructuring thus specified can be summarized as a set of objectives:

Table 3.1 Objectives for a sustainable restructuring of the energy system

Objectives	*Concretization*
Availability of resources	Period of secure practice
Energy system	Reliability (end user), openness of options, risk avoidance
Environment	Climate change, emissions, surface consumption

Whenever in this study we point out the necessity of a sustainable restructuring of the energy system, one has to keep in mind all of the goals cited above. However, for this set of goals, the reduction of CO_2 emissions can serve as a "guiding indicator" that is supplemented by further indicators (land use, openness of options etc.) in a particular context.

Facing the knowledge deficits discussed above, environmental policy usually calls upon the *precautionary principle*[5], according to which measures are taken as soon as scientific plausibility is given, in the sense of a "potential for concern". This *plausibility* condition may be met by a climate change caused by emissions of climate-effective gases. The measures to be taken, however, shall be chosen according to the principle of proportionality (prohibition of excess). Again, there is no point of reference sufficiently precise for a balanced consideration i.e. for the comparison between expected costs and damages incurred by different strategies. The now common form of the precautionary principle, the ALARA principle ("As Low As Reasonably Achievable") highlights this observation insofar as the assessment of what is "reasonably achievable" is just not possible. Nevertheless, unswerving insistence on the principle of proportionality appears to cancel out the precautionary principle.[6]

In principle, if we accept a however weak potential for concern, counteraction appears to be advisable. Still, for the reasons mentioned above, it is difficult to determine the kind and extent of such counteraction. Goals related to resource and energy economy are needed to serve as "concretization aids" that allow the determination of the proportionality of the measures in question (see chapter 4). In any case, independent of the weight of the different arguments, the discussion about the restructuring of the energy system requires us to know the medium- and long-term costs and benefits of a sustainable restructuring driven by innovations.

---

[5]   See Rehbinder 1991; id. 1998, *Rdnr* 27ff.
[6]   Bringing forward the precautionary measures is regarded as justified in cases where "severe, possibly irreversible" effects are at issue (Rehbinder 1998).

3.2
Sustainable development and justice

3.2.1
Introduction

Sustainability is a normative idea, without any doubt, and each of the different concepts introduced in chapter 2 involves a specific normative decision of general principle, which in turn involves specific ethical considerations in each case, even if they are not always explicitly mentioned. From our point of view, the position that there are no obligations with regard to future generations is hardly advocated in the somewhat sarcastic question ironically proposed by one of the "Marx Brothers", Groucho Marx (*„ Why should I care for posterity? What has posterity ever done for me?"* [7]). But still we found similar views in the German business ethics discussion, for instance, albeit only in the mild form that the interests of future generations are to be considered today only to the extent that the present generation benefits from it as well (Karl Homann). Consequently, limiting the opportunities for activities of the present generation is only admissible if this generation is better off (or at least not worse) in this case than in a situation without such limitations (Homan 1996, p. 42 ff.). The obvious problem with this position is that whenever there is no room left for Pareto improvements [8], i.e. in every instance of a true trade-off situation and hence a grave ethical conflict, it gives us no rules or recommendations at all. [9] Therefore, applying an intertemporal Pareto criterion can only be a first and ethically largely abstinent step towards answering the questions at hand. If intertemporal Pareto improvements are not (or no longer) possible, one has to look for justifications for the necessary trade-offs (balances) between different options.

Whether there are opportunities for Pareto improvements in specific questions of intergenerational distribution inherent in the sustainability discussion, as in the case of sustainable energy innovations discussed in this book, or whether there are, at least in some cases, clear-cut trade-off situations, this can only be answered on an empirical basis in each concrete case. However, any conception of sustainability introduces some type of intertemporal stock. The requirement to conserve it – or the assumption that this would be fair – puts the concept of sustainability into a normative context.

In the relevant political documents (e.g. WCED 1987), the sequence of, first, introducing a concept and, second, using it in a normative way is carried out in a single step. Still, a more differentiated consideration is helped by treating the two

---

[7]   Quoted from Vallance (1995, p. 115).
[8]   Pareto improvements are such improvements by which the situation of an individual member of a society can be enhanced without worsening the situation for any other member of the society. A situation where such improvements are not possible anymore is referred to as a "Pareto optimum" (Sohmen 1976).
[9]   As Amartya Sen (1987, p. 35) pointed out: „Despite of its general importance, the ethical content of this welfare economic result is, however, rather modest. The criterion of Pareto optimality is an extremely limited way of assessing social achievement [...]." In this, Professor Sen refers to the first and second laws of welfare economics, which show the equivalence, under very specific conditions, of the Pareto optimum and a general balance of competition.

steps of the sequence as two separate entities. The first step was discussed in the previous chapter, where we indicated its implications for the second step. Now we will point out, with regard to the second step, how moral intuition allows us to regard questions of justice as uncontroversial – apart from the details – even in an intergenerational sense. Practically every human being will accept long-term obligations, at least towards the immediately following generations. In a similar way, a decreasing commitment of those obligations with increasing distance in time (gradation) will also find intuitive consent because one will be more inclined to bequeath a certain advantage to his own children (which one normally knows) than to his descendents in the 10th generation (which one cannot even imagine). In addition, this gradation of obligations can be justified by means of the increasing uncertainty about the occurrence of desired impacts of action as time goes by (Gethmann/Kamp 2000).

In our notion of critical sustainability, a certain idea of intergenerational justice is implied insofar as the standards which have been deemed critical ought to be maintained. Critical sustainability as a dynamic concept for using resources is involved in issues of gradation, especially with regard to the increasing uncertainty of our specific knowledge in the course of time. With this, the context for the following discussion is outlined completely. Hence there is no need, at this point, for going into the many branches of the philosophical and economic discussion about justice and, especially, for determing how the gradation of commitments could be shaped.

3.2.2
Political Approaches

One can say that the principle of intergenerational justice or equity underlies the WCED view of sustainable development, which

> 'aims to ensure that economic progress [and we should add, development in general; WA] does not prejudice the chances of future generations by depleting the natural capital stock that sustains human life on the planet. Equity requires that this strategy is followed by all societies, both rich and poor' (CGG, p. 52).

In other words, the Brundtland view orders two problems of justice: justice between North and South (rich and poor, i.e. intragenerational justice) can and should be pursued but not in a way that harms the pursuit of intergerational justice (Hengel 1998). Thus, intergenerational justice is seen here as a moral constraint on the pursuit of intragenerational justice. The content of the constraint is that generations should have equal opportunities to use at the very least primary natural resources (in a broad sense) or perhaps, adopting a more recent concept, environmental (or ecological) space. As the elementary but very broadly shared moral basis of the strategy of sustainable development one can take the golden rule (Rawls; CGG p. 49): people should (not) treat others as they would (not) themselves wish to be treated. So far the relation between sustainability and justice may seem obvious but also rather stipulative. A more internal relation might strengthen the case for both of them. This also requires elaborating a more specific conception of e.g. intergenerational justice. So let us see what a more internal and specific account may look like. To begin with, it should meet some general requirements having mostly to do with the viewpoint of legitimacy. Because in pursuing sustainable development the rich countries should lead by example, it is appropriate that the desired account is based on a liberal, particularly a liberal-egalitarian theory of justice. A theory, moreover, which is also universalist in scope and so in its application spatio-temporal neutral. Both characteristics seem appropriate if the theatre of operations for the theory comprises liberal democracies but in a rapidly globalizing context and in a long-term perspective. Interesting examples would be Rawls' political liberalism (with its revised 'just savings' principle), Barry's account of (intergenerational) justice; Wissenburg's 'restraint' principle and Norton's account of intergenerational equity.

Because the accounts of justice mentioned imply mostly a rather minimalist view of our moral responsibility for or toward future generations, it may be helpful to conceive them in the framework of 'Praxisnormen' proposed by Dieter Birnbacher (1988, p. 197–240). Although he justifies them in a indirectly utilitarian way, they can in this context be adopted equally well by adherents of other ethical orientations, especially because they have a close affinity with the development-aspect of sustainable development Birnbacher specifies the following 'Praxisnormen', which are norms for collective action in particular. Human action should *not* be such as to a. threaten human survival and the survival of higher species (i.e. collective self-preservation); b. put in jeopardy a decent human existence in the future (broad no harm principle). Moreover, we should take care to c. avoid creating additional irreversible risks; d. preserve and enhance present natural and cultural resources; e. help others in the pursuit of these and other future-oriented goals; and f. educate the next generation(s) in the spirit of these norms.

3.2.3
The theory of justice (Barry)

To flesh out the so far rather abstract remarks about justice and justification, it may be helpful to see what an account of intergenerational justice may look like. *Brian Barry*'s account seems an attractive one because it ties in explicitly with the concern about sustainability.

Barry (all references are to his 1999 paper unless otherwise indicated) starts with a question about the 'ethical status' of sustainability: 'Is sustainability ... either a necessary or a sufficient condition of intergenerational distributive justice?' He takes justice here in a narrow sense which focuses on 'conflicts of interests', so that questions of intergenerational justice are 'characteristically questions of intergenerational distributive justice'. In answering his question he starts from a premiss of the fundamental equality of human beings. From this premiss, taken by him as an axiom, he derives four theorems.

'1. *Equal rights*. Prima facie, civil and political rights must be equal. Exceptions can be justified only if they would receive the well-informed assent pf those who would be allocated diminished rights compared with others.

2. *Responsibility*. A legitimate origin of different outcomes for different people is that they have made different voluntary choices. (However, this principle

comes into operation fully only against a background of a just systems of rights, resources and opportunities.) The obverse of the principle is that bad outcomes for which somebody is not responsible provide a prima-facie case for compensation.

3. *Vital interests.* There are certain objective requirements for human beings to be able to live healthy lives, raise families, work at full capacity, and take part in social and political life. Justice requires that a higher priority should be given to ensuring that all human beings have the menas to satisfy these vial interests than to satisfy other desires.

4. *Mutual advantage.* A secondary principle of justice is that, if everyone stands *ex ante* to gain from a departure of a state of affairs produced by the implementation above three principles, it is compatible with justice to make the change. (However, it is not unjust not to.)' (p. 97–98)

Obviously, mutual advantage (a Pareto improvement) which does not derive from such a departure, e.g. increased energy efficiency, is a fortiori compatible with justice and is in the interest of all concerned. Barry has doubts about this as far as the distant future is concerned because such improvements are expressed in terms of preferences and to know these becomes more difficult the further we have to look into the future. But what other implications do these principles have for intergenerational justice? The first, equal rights, has no direct intergenerational implications. The same goes for the second, which is applicable foremost within generations, unless we take the view that people can control the size of the population (a topic that will not be discussed here) or unless we can foresee that because of our use of resources in the present we leave future generations less than we started with. In the latter case (above all the use of non-renewable resources) we have the duty, of course, to leave them compensating "assets", e. g. additional productive capacity. It is the third principle, vital interests, which has direct consequences; this is by way of the universalist idea that 'locations in space and time do not in themselves affect legitimate claims' (p. 99; 100). The implication is that the vital interests of people in the future have the same moral weight as similar interests in the present.

The core concept of sustainability, which is 'irreducibly normative' (p. 105), is, as Barry puts it, that there is 'some X whose value should be maintained, in as far it lies within our power to do so, into the indefinite future' (p. 100). The content of X should not be utility in the sense of preference-satisfaction, Barry argues, but the chance of members of future generations to live the good life according to their own conceptions of it (which do not exclude our conceptions of it). So X needs to be conceived as 'some notion of equal opportunity across generations' (p. 104). And this has to be conceived, according to Barry, as maintaining conditions such as 'to sustain a range of possible conceptions of the good life.' Harming their vital interests is not a part of these conditions but neither is leaving them a world in which nature is 'utterly subordinated to the pursuit of consumer-satisfaction' (p. 105). We may see here a moral basis for Birnbacher's important norm of not creating irreversible risks (see already Barry 1978, p. 243; 1977, p. 275; and, very clear, Goodin, about 'keeping the options open' in 1982, p. 209 ff.).

It follows from these considerations, especially from the principle of responsibility, that sustainability is at least a necessary condition of intergenerational justice. For 'no generation can be held responsible for the state of the planet it inherits' (p. 106). So, to begin with they should not be worse off than we are. But 'the potential for sustaining the same level of X as we enjoy depends on each successive generation playing its part. All we can do is leave open the possibility, and that is what we are obliged in justice to do' (p. 106).

So far the argument has been based upon the presumption of a fixed population. If we take into account population growth and, moreover, that the X which has to be maintained is defined over individuals, matters are becoming much more complicated and the prospects of sustainability much less bright. There is not much the present generation can do about the size and growth of the present population, but it can be held, at least to a limited extent, responsible for them. To the extent we take this responsibility we will be more just to our successors (leaving them less resources spread over fewer people) than otherwise (p. 111). If, for the case of population growth, we subordinate the principle of vital interests to the principle of responsibility, we might even say that that sustainability is even a sufficient condition of intergenerational justice: what more could we reasonably be expected to do?' (p. 112).

Of course, we have also to take into account – for 'the principle of vital interests forces us to focus on the fates of individuals' (p. 112) – that the vital interests of many people in the present are not met and that the vital interests of many people in the future will predictably not be met too, even if the criterion of sustainability is satisfied. This suggests, according to Barry, that there is not an absolute distinction between intergenerational and intragenerational justice. One might, again, here invoke the principle of responsibility as having priority above that of vital interests, but we 'must recognize that intragenerational injustice in the future is the almost inevitable consequence of intragenerational injustice in the present' (p. 113). All the more reason, one is inclined to say, to try seriously to meet the vital interests of people now living without putting sustainability in jeopardy.

A certain tension in Barry's attempt to justify sustainability in a liberal-egalitarian framework should by now be clear. As he argues, sustainability obliges us to think in terms of (conceptions of) the good life: the possibility of the good life in the future is what is at stake in the pursuit of sustainability. But policy in liberal democracy is supposed to be neutral in respect to conceptions of the good life! So taking sustainability seriously makes it unavoidable to rethink and reappraise the nature and the boundaries of liberal neutrality. We will meet this theme again in the discussion of 'weak' versus 'strong' conceptions of sustainability and in the debate on sustainable patterns of consumption. Particularly changing patterns of consumption may prove to be very challenging within a liberal framework of thought. For consider that present-day affluent societies manifest a 'widespread, shared belief in the value of consumption for everyone, a belief that what can be achieved through consumption is at least part of "the good life for humans", and hence that "the good society" is one that provides ample opportunities for people to enjoy these benefits, and indeed to an ever-increasing extent' (Keat p. 342). So, infringements on the liberty to consume or even a decline in real income (on account of price rises while incomes remain the same) may cause much resentment among consumers and thus bodes not well for environmental policies aiming at reduction of consumption, even if it means only (!) reduction of the material and energy content of current patterns of consumption (in case increases in efficiency and substitutions are not enough to bring that reduction about).

3.2.4
Vulnerability, the Future and the Environment (Goodin)

The emphasis so far has been on justice and its theoretical elaboration which is not surprising since justice is a core idea in the predominant conceptions of sustainable development. This does not mean there are no other ethical ideas on which to base our moral responsibility for future generations, the poor and the environment (which are implicit in any acceptable account of sustainable development). It may be enlightening and morally attractive to some readers to briefly discuss another ethical approach (informed by utilitarianism) which emphasizes the moral relevance of vulnerability. At any rate this approach, developed by Goodin, would seem to embody an apt moral view of our relations to the poor and future generations. Additionally, this approach also leads naturally to a global and long-term perspective on the environmental plight of liberal democracies. In what follows the emphasis is on future generations.

Future generations are especially vulnerable to the harmful effects of our choices and behaviour. They are heavily dependent on us for the material and cultural resources available to them. According to Goodin, our responsibility to leave them enough resources has its strongest moral basis in this vulnerablity. The case for this view consists of a negative and a positive part. The negative argument amounts to a rejection of theories of intergenerational justice, especially tot the extent they imply reciprocity between us and members of future generations. The positive argument invokes special duties, duties of the type parents have towards their children (sometime the reverse too), spouses or friends towards each other; duties to help needy strangers, if one can help without too great sacrifices to oneself and if others are not around or able to help, also belong here. Crucial to special duties are, in Goodin's view, aiding and protecting dependent and therefore vulnerable people. If we take this view, it becomes readily clear that these special duties are not so special after all, because vulnerability by dependency if rather frequent in our society. This leads to a *general positive* duty to help others in vulnerable positions.

The account of sustainability one could elaborate from these and similar suggestions need not be limited just to the idea of maintaining (human) welfare or utility over time with environmental concern in a subordinate position. Goodin himself has in the meantime extended the theory outlined to a consequentialist theory of value of nature which can be used to argue for strong policies for the protection of nature and the global environment (Goodin 1992 and, especially, 1996).

3.3
Efficiency and sufficiency – discussing sustainability in theoretical and practical terms

As explained at the beginning, the purpose of the present study is to explore the potential of innovations for a sustainable restructuring of the energy system, and to propose specific measures for exploiting and, eventually, realizing these potentials and strategies.

Apart from developing and establishing more efficient means (efficiency improvements), it is of course possible to modify the ends that are to be achieved by these means. For the purposes of discussing and transforming these ends, the rather unfortunate term "sufficiency" is used quite frequently. The term is unfortunate because it suggests starting from a catalogue of means, compiled according to criteria of sustainability, and then discriminating between constellations of ends depending on whether those means are sufficient for achieving the ends. Herman Daly, who originated the concept of the extent of using load capacities ("scale") in the sufficiency discussion, chooses the allegory of a boat on which goods are transported (Daly 1996, p. 50). If one fails to ensure that the goods are evenly distributed on the loading area, the boat can overturn and sink. This mistake can be avoided by improved loading (efficiency); but the load line of the boat, which represents the load capacity, cannot be exceeded without sinking the boat (sufficiency). However, the load mark cannot be established with sufficient precision, because of the uncertainty of the available knowledge, as explained above. Therefore, this concept of sustainability merely helps to distinguish between more or less sustainable options instead of formulating a precise line as the limit of sustainability. The decisive criterion is not the absolute weight, but the decrease in weight of the goods.

In the study *"Zukunftsfähiges Deutschland"* ("Sustainable Germany", BUND/ Misereor 1996, p. 218), in particular, efficiency improvements as well as the idea of "sufficiency" are emphasized. The study highlights not only the possibility of meeting human needs on the consumption side (not only with regard to the technical efficiency of the production and use of goods) using lower inputs of energy, resources and materials, guided by the criteria of "economizing, regional orientation, shared use [of long-lived consumer goods], durability"; it also stresses the idea that such changes should not be seen as asking for sacrifices and restrictions but rather as some kind of "ecological progress" leading to both the re-achievement of "wealth in time instead of wealth in goods" (p. 221) and to a new "elegance of simplicity" (p. 223). The examples of such a new lifestyle cited there may be vivid, but the study does not prove the general applicability of those welfare-improving sufficiency strategies for all groups of consumers. In reality, sufficiency can and will still involve restrictions and sacrifices in many cases. Options like that cannot be decreed from above in democratic societies. Ultimately, they can only be demanded and enforced by the people themselves.

In the light of these considerations, one can assess whether, and to what extent, efforts to transform ends make the existing energy system more sustainable. Following the distinction between efficiency (in the production and use of goods) and sufficiency (in the concrete shaping of lifestyles), one can more generally refer to a *theoretical* and a *practical* discussion of sustainability[10]. The theoretical discussion is concerned with improving the means for achieving given ends; the practical debate deals with the examination and transformation of these ends. In the following, we will mainly discuss theoretical aspects, whereas practical aspects are an issue especially in the sense that political instruments that change price ratios intervene in every individual's potential for achieving his or her objectives. To a certain extent, such measures are accepted as options of government action.

As already indicated, the scope of political measures is narrowly limited when such action aims at a far-reaching transformation of ends; it could only be widened by running the risk of state paternalism. Therefore, political measures should be facilitating, but not restrictive in character. To reduce the energy consumed by commuting between home and work place, for instance, limiting the commuting distance through statutory regulation could be taken into consideration as a restrictive measure. Such legislation would, however, be bound to fail, if only for constitutional reasons. Instead, one could support a new thinking in town planning, which as such would lead to a fusion of residential and commercial areas, or one could aim at an improvement of local and regional public transport systems. Still, state activities are not the only option. The transformation of ends can lead to a more sustainable energy system as a result of cultural developments, too. However, with regard to these types of transformations, we cannot formulate any recommendations for governmental actions, apart from providing information e.g. through appropriate product labeling (see chapter 7).

3.4
Interim conclusions

The considerations so far lead to some important findings, which will be further scrutinized in the following chapters. It emerges that there is a multitude of relationships between "innovation" and "sustainable development" – both concerning the methods and, above all, the substance –, which can be treated paradigmatically for the energy sector. On the one hand, society and economy depend crucially on the availability of energy; on the other, the goals defined in chapter 3.1.2 cannot be achieved in a lasting way by perpetuating the status quo. Therefore it seems that significant changes in "economy-style" – not least concerning energy consumption, which is supported, in its bulk, by using non-renewable energy carriers – cannot be dismissed. Without sustainable innovations, such changes stand virtually no chance of success.

Quite generally, innovations oriented towards sustainable development can (and should) lead to a situation where economic and social achievements regarded as necessary can be produced with a lower "consumption of nature" than would be possible in the absence of such innovations. This implies the requirement – for which we have already argued – to develop the concept of sustainability from a (static) inventorial concept into a (dynamic) concept of utilization. According to the

---

[10] In this application, the terms "theoretical" and "practical" refer to different modes of philosophy. Theoretical philosophy includes epistemology, for instance, while ethics is part of practical philosophy.

latter, technical progress, social and organizational change etc. are possible contributions of the generations living today to protect the interests of future generations, even in cases (e.g. exhaustible energy resources) where the preservation of stocks, in the classical sense, is not possible.

The emphasis on environmental and resource issues does not imply a general rejection of the three-pillar model of ecological, economic and social sustainability often argued for, since the latter takes account of the fact that ecological objectives can only be aimed at successfully if the economic performance (on the national, transnational and global scale) is not diminished substantially, and if dangerous social distortions (within and between societies) are avoided. Yet it takes note of a certain asymmetry between ecological postulates concerning sustainability, on the one hand, and economic and social sustainability demands, on the other: In as far as the first are based on laws of nature, they must always be considered with the highest priority; for such laws cannot be influenced, either by political, or by educational or information efforts. Again, this statement must be qualified by the observation that, contrary to common conceptions, there is no "reference work on the laws of nature" where those rules can be looked up. As we discussed above, the uncertainty of scientific results is constitutive especially in the areas that are relevant for the debate about sustainability.

In contrast to the economic and social "pillars", requirements for preserving nature have no voice in the political process, which is why the ecological dimensions are often neglected in favor of special interests and avoidance of adjustments and social hardship. Ecology is an "external effect" not only for the market but also for politics. This is particularly true if the effects only emerge after a long time, are scientifically uncertain or controversial, or not immediately "noticeable" (as was the case of air pollution, for instance, in the past). Science can, and must, act as a corrective here by emphasizing the neglected ecological dimension. Hence, sustainable energy innovations are defined, in the present study, as new technological, organizational or institutional solutions or processes leading to a qualitative improvement of the environment and a reduction of environmental risks, without inducing unacceptable economic or social disadvantages. This is achieved mainly by keeping such disadvantages as small as possible and compensating them by additional measures, if necessary.

Although it may be possible to change social habits through information and education efforts, innovations that influence the functioning of economic and social subsystems seem more promising for promoting sustainability than those trying to influence the thinking of the individuals concerned. For this reason, "economic instruments" of environmental protection are of particular importance in all innovations oriented towards sustainability. Simply through an ecological adjustment of price ratios, they not only induce us to use existing technologies in a manner as environmentally sound as possible; they also provide powerful incentives to give the innovation activity – whose substance cannot be predicted in any concrete detail – generally a more "environment-friendly" direction.

Following these foundations, the next step, taking into account the results worked out so far, will be to analyze the sustainability of the (future) energy system and the (technical) potentials for sustainable energy innovations (increasing energy efficiency and developing regenerative energy sources).

4 Towards a Sustainable Energy System – Legal Basis, Deficits and Points of Reference

4.1
Fundamental legal standards for a sustainable energy system

As explained in the introduction, the research group does not set its own objectives and standards on which the analysis is based; but selects its criteria from legal standards set out in international treaties and from EU and national law.[1] Naturally, such legal standards do not represent (quantitative) targets that give rise, conclusively, to specific strategies. However, they do define – as a "regulative concept of sustainability", so to speak – directions and limits, within which politicians and other agents have to set concrete objectives and develop programs (with sustainable development as the constitutive concept). The conceptional considerations developed in chapters 2 and 3, as well as the criteria for evaluating and decision-making, help execute the transition between the "abstract legal standard" to concrete recommendations for action in a transparent way. They are instrumental not only for concretizing, but also for developing further those standards without claiming to be comprehensive or even aiming at evaluating the consequences of the sustainability concept for the legal system.

4.1.1
Developments in international law concerning climate protection

The Rio Conference

Following its introduction by the Brundtland Commission, the concept of sustainable development was taken up and further elaborated in the documents of the United Nations Conference on Environment and Development (UNCED) in Rio de Janeiro, most notably in Agenda 21. "Protection of the atmosphere" is highlighted especially in chapter 9 of Agenda 21, even if it does not put the signatories under any obligation. The United Nations Framework Convention on Climate Change (UNFCCC), on the other hand, was the first definitive document on climate policy to be legally binding in terms of international law. It was developed by an intergovernmental negotiating committee appointed by the UN (INCC[2]), signed by about

---

[1] It is not our intention to give a comprehensive review of the sprawling legal discussion on the subject of sustainability; we will rather discuss some aspects that are important with regard to the recommendations for action presented in chapter 7.

[2] Intergovernmental Negotiating Committee for a Convention on Climate Change.

160 countries at UNCED, and has been in force since March 21, 1994.[3] According to article 2, the main objective of the UNFCCC is the "stabilization of greenhouse gas concentrations in the atmosphere at a level that would prevent dangerous anthropogenic interference with the climate system". This means that the aim is, merely, to limit the increase of greenhouse gases in the atmosphere.[4] Article 3 of the UNFCCC takes up specific components of the principle of sustainable development, by referring to the precautionary principle (par. 3), and by the stipulating that the climate system should be protected for the benefit of present and future generations (par. 1). In art. 3, par. 5 of the UNFCCC, the objective of a "sustainable [...] development in all parties [to the convention]" is mentioned explicitly.

Specifically for the energy sector, the following statements are most relevant: According to art. 4, par. 2 (a) UNFCCC, each of the so-called developed countries listed in Annex I binds itself to "adopt national policies and take corresponding measures on the mitigation of climate change, by limiting anthropogenic emissions of greenhouse gases and protecting and enhancing its greenhouse gas sinks and reservoirs". Art. 4, par. 2 (b) UNFCCC stipulates that the signatories must report periodically to the conference of the parties to the convention, with the aim, to be achieved by the year 2000, of "returning individually or jointly to their 1990 levels these anthropogenic emissions of carbon dioxide and other greenhouse gases not controlled by the Montreal Protocol".[5] However, these political commitments are still[6] not binding with regard to international law,[7] although they contain the basic elements of the international understanding of the concept of sustainable development in the energy sector.

From the legal documents of the Rio Conference it was deduced that the exploitation of non-renewable resources should be limited to such materials for which there are long-term replacements in the form of renewable primary resources, or of secondary resources.[8] Yet, in this form, the so-called Second Rule of Management is not laid down in international law; the phrase "long-term" makes it hardly practical, due to unforeseeable developments concerning the economic usefulness of additional resources (e.g. oil shale in the place of crude oil); furthermore, the needs of present generations,[9] which also exist and must be equally respected in the context of sustainable development, would not be met if the rule were to be applied strictly.[10]

---

[3]   *Secretariat of the UNFCCC* (ed.), United Nations Framework Convention on climate Change, p. 1.

[4]   See also Hohmann, *Ergebnisse des Erdgipfels von Rio*, NVwZ 1993, 311/316.

[5]   V. 1987-9-16, BGBl. II 1988, p. 1015. This protocol mainly aims at protecting the ozone layer by gradually outlawing the use of CFCs.

[6]   See art. 25 para. 1 Kyoto Protocol. On the pending ratification, ibid. under 2.

[7]   See Kloepfer, *Umweltrecht*, 2. edition 1998, § 9 Rn. 84; Bail, *Das Klimaschutzregime nach Kyoto*, EuZW 1998, 458.

[8]   Bundesministerium für Umwelt, Naturschutz und Reaktorsicherheit, *Bericht der Bundesregierung anlässlich der VN-Sondergeneralversammlung über Umwelt und Entwicklung 1997 in New York, Auf dem Weg zu einer nachhaltigen Entwicklung in Deutschland*, BT-Drucks. 13/7054, p. 6, and Enquête Commission *"Schutz des Menschen und der Umwelt"*, BT-Drucks. 13/7400, p. 13.

[9]   A "development that meets the needs of the present without compromising the ability of future generations to meet their own needs", World Commission on Environment and Development, *Our Common Future*, 1987, p. 43.

[10]   For details, see Frenz, *Sustainable Development durch Raumplanung*, 2000, p. 20 ff.

The Kyoto Protocol

Basic duties. Following the Rio Conference, international climate policy has been valorized significantly. On the occasion of the third convention of the Conference of Parties (COP-3)[11] to the UNFCCC in Kyoto on December 12, 1997, the so-called Kyoto Protocol was adopted, which, in contrast to the Framework Convention itself, set legally binding[12] emission targets aiming at the reduction of CO_2 emissions from the industrial and transformation countries. Of course, according to art. 25, par. 1, the Kyoto Protocol will become legally binding under international law only when at least 55 signatory countries, or their parliaments, have ratified the protocol and deposited the corresponding deed of ratification with the Secretary General of the United Nations. Furthermore, among those countries having ratified the protocol must be Annex-I countries[13] that are jointly responsible for at least 55 % of the CO_2 emissions in the year 1990, which are cited in this annex. Not one of the signatory parties listed in Annex I of the Kyoto Protocol had deposited a deed of ratification by the end of 2001. Since then, however, such documents have been submitted by the European Union or its member states, respectively.

The commitment period laid down in the Kyoto Protocol – i.e. the years in which, on average, the goals for the individual countries, as defined in Annex B, must be achieved – is five years (2008-2012), according to art. 3, par. 1. The central stipulation is laid down in article 3, by which the industrialized nations included in Annex I commit themselves not to exceed – individually or jointly[14] – their assigned emission limits or to default on the country-specific reduction commitments, respectively, inscribed in Annex B. The aim is to reduce the overall emissions of greenhouse gases (GHGs) by all the industrialized countries listed in Annex I by at least 5 % of the 1990 levels.[15] GHGs include CO_2, methane, nitrogen oxides, chlorofluorocarbons (CFCs) and some others; only the first three gases are relevant for the energy sector. All GHGs are converted into so-called CO_2 equivalents, according to their effects on the climate. The GHG emissions stemming from the provision and use of energy account for ca. 80 % of all anthropogenic emissions. Nearly 95 % of these energy-related emissions is CO_2, about 4 % is methane; the remainder are nitrogen oxides. According to Annex B, all member states of the EC have committed themselves to a reduction of 8 % compared to the situation in 1990.[16] The individual reduction commitments vary considerably among the EU member states, between – 28 % for Luxembourg and + 27 % for Portugal. The target for Germany was set at – 21 %. According to art. 3, par. 3, this number may reflect not only the reduction in emissions, but also the removal of greenhouse gases through increased CO_2 absorption, which can be achieved by land-use changes and forestry activities.

---

[11] On details of the development in climate policy between the environment summit of 1992 in Rio de Janeiro and the 3rd Conference of the Parties (Kyoto 1997), see Ehrmann, *Die Genfer Klimaverhandlungen*, NVwZ 1997, 874 ff.

[12] Bail, *Das Klimaschutzregime nach Kyoto*, EuZW 1998, 460.

[13] Annex-I states are the industrial countries listed in Annex I of the Framework Convention on Climate Change.

[14] Referring to joint implementation of policies and measures.

[15] Art. 3, par. 1.

[16] USA: - 7 %; Japan, Canada, Poland and Hungary: - 6 %; Croatia: - 5 %; New Zealand, Russia, Ukraine: 0 %; Norway: + 1 %; Australia: + 8 %; Iceland: +10 %.

The Kyoto Protocol declares climate protection to be a core element of sustainable development. With regard to Annex-I countries realizing their targets, one has to distinguish between national and international measures. The exemplary list under art. 2, par. 1 (a) may serve as a (albeit rudimentary) guide to national policies and activities that would support a sustainable development. The "enhancement of energy efficiency", "research on, and promotion, development and increased use of, new and renewable forms of energy [...] and innovative environmentally sound technologies" and the limitation and/or reduction of methane emissions in connection with the production and distribution of energy are the propositions of particular relevance for the energy sector. According to art. 2, par. 1 (b), the signatory parties are urged to cooperate in order to enhance the effectiveness of the individual policies and measures by exchanging their experiences and information.

Flexibilities. The international mechanisms are meant primarily to support economically efficient ways of climate protection. The aim is to reduce global greenhouse gas emissions at optimum cost efficiency. To this end, the Kyoto Protocol provides various new mechanisms under international law.

The mechanism for environmentally sound development provided in the protocol (art. 3, par. 12 in conjunction with art. 12) allows the Annex-I countries to carry out emission-reducing projects in developing countries and to use the emission reductions certified as such for [meeting] their own obligations concerning limitations and reductions. According to art. 12, par. 2, this mechanism serves "to assist Parties not included in Annex I in achieving sustainable development and ... to assist Parties included in Annex I in achieving compliance with their quantified emission limitation and reduction commitments under article 3". Hence, the goal is to enable a cost-efficient realization of the reduction targets on the part of the industrial countries, on the one hand, and to promote the cooperation or technology transfer between industrial countries and developing countries, on the other. However, according to art. 3, par. 3 (b), the industrial countries are allowed to meet only part of their commitments concerning emissions and reductions by way of emission reductions through clean-development projects.

Another way of fulfilling the legal commitments concerning reductions [by activities] on foreign territory, though limited to the territories of the Annex-I countries, is opened up by the option of joint project implementation according to art. 3, paragraphs 10 and 11 in conjunction with art. 6, par. 1. Accordingly, the "reductions account" of a country in western Europe, for instance, might be credited with the emission reductions following from emission-effective modernizing measures at outdated, eastern European power plants. However, according to art. 6, par. 1 (d), the reduction units acquired in this way are only allowed to supplement domestic measures. Furthermore, art. 6 only covers reduction units stemming from concrete emission-reducing measures.

Emission trading among industrial countries also enables a shift of commitments. In fact, no real efforts have to be made insofar as rights are purchased from a country that keeps within the agreed emission limits and capitalizes on this situation by selling certificates instead of crediting itself with over-fulfillment beyond its obligations. Still, by stipulating that the trade in emission rights shall also be no more than "supplemental" to any domestic measures, art. 17 presumes an upper limit on such dealings. The interpretation of this element of fact is extremely controversial. While the US insisted on full flexibility, even before their complete withdrawal from the Kyoto Protocol under President Bush, the EU argued that half of the reduction commitments must be fulfilled domestically.[17] "Supplemental" does not mean "exclusive"; but implies a mere extra function accounting for 50 %,[18] if not 49 % as an upper limit, if the majority contribution is to come from domestic measures and only an additional, supplemental part is to be contributed through measures in another country, as the wording of the Kyoto Protocol suggests.

Beyond that, art. 17 demands that the "principles, modalities, rules and guidelines, in particular for verification, reporting and accountability" are defined and elaborated by the Conference of the Parties.

Continuation at The Hague, Bonn and Marrakesh. These and other questions still open after Kyoto were supposed to be resolved at the 6th Conference of the Parties (COP-6) at The Hague, which convened from November 13-24, 2000 – and ended in failure. The main obstacle to a common solution was disagreement about the question to what extent a reduction, or storage, of greenhouse gases through so-called sinks created by land-use changes and forestry measures should be taken into account with regard to the reduction duties of the signatory parties. The reluctant position of the European Union is based on the fact that, presently, there is little scientific evidence for carbon sinks being allowable in the context of emissions; but closer inspection does reveal a multitude of scientific questions. In this context, we should point to "Background Information" published by the Max Planck Society for the Advancement of Science (Germany), about the role of forests and their cultivation in the natural carbon cycle and their effects on the global climate. There it is stated that "presently, science does not know exactly yet which ecosystems are so-called carbon sinks, whether these sinks are sustainable, and how the enormous amounts of carbon stored in soils should be assessed". At the same time, the contest is underway to decide which region is a global sink for, and which is a global source of CO_2. For instance, one publication from North America shows that the USA is the only carbon sink in the northern hemisphere. Since then, another publication has stated the conflicting position that Europe is the only sink among the countries of the northern hemisphere."[19]

The negotiations, "interrupted" in The Hague, were resumed in Bonn in July 2001. At that point, the US had completely rejected the Kyoto protocol. The EU still reached a compromise with Japan, Canada and Russia, so that the protocol could be ratified regardless of the rule that 55 countries must ratify, and that these countries have to represent 55 % of the emissions of the industrial countries, with the USA contributing 40 %. This agreement, however, came at the price of a far-reaching erosion in content: Biological sinks for storing carbon monoxide shall be allowable to a large extent and on varied terms depending on the country concerned. This particularly favors the three parties to the compromise and reduces the obligations of

---

[17] See also: *FAZ* of 11-18-2000, no. 269, p. 15, right column, *"Nur ein Tropfen auf den heißen Stein"*; *FAZ* of 11-23-2000, no. 273, p. 19 *"BDI: Klimavereinbarung ist zwingend nötig"*.

[18] Müller-Kraenner, *Zur Umsetzung und Weiterentwicklung des Kioto-Protokolls*, ZUR 1998, 113/114: upper limit of 50 %.

[19] *Max-Planck-Gesellschaft zur Förderung der Wissenschaften e.V.*, http://www.mpg.de/pri99/hg_hainich.htm.

the industrial nations by ca. 169 million tons of CO_2. The so-called flexible mechanism, with its provisions for accounting extraterritorial activities against obligations to reduce emissions domestically, may be applied in a generous fashion, even if nuclear power plants must not be taken into account. Developing countries, including the OPEC states, receive subsidies of up to $ 410 mil. from the EU and other industrial countries to support climate protection projects.[20] Strict result checking in the form of legally binding sanction mechanisms in the case of noncompliance with reduction obligations, which was not part of the agreement, was the subject of a follow-up conference in Marrakesh, held from October 29 to November 9, 2001.

At the Marrakesh conference, more detailed reporting duties and a so-called compliance system were agreed, still leaving open the question whether or not that system is binding under international law. An "enforcement branch", composed of six representatives from developing countries and four from industrial nations, decides on noncompliance with obligations arising from the agreement. The same body establishes an action plan to be implemented where emission and reporting duties are not fulfilled. Countries that fail to meet their reduction obligations are excluded from selling their emission allowances to other signatory parties.

Outlook. Considering the close connection between climate protection and the postulate of sustainability in international law (concerning the environment) formulated in the Kyoto Protocol, and taking into account the flexible mechanisms proposed as a problem-solving strategy as well as the domestic policies and measures cited as examples therein, we are confronted with the question to what extent these tools are appropriate or sufficient to achieve the long-term objective, laid down in art. 2 UNFCCC, of preventing dangerous, anthropogenic disorders of the climate system. This question must be answered especially with regard to the responsibility of generations living today for future generations, the intergenerational component of sustainability, which is generally acknowledged to be the central idea of "sustainable development". The idea is taken up in art. 3, par. 1 UNFCCC: "The Parties should protect the climate system for the benefit of present and future generations of humankind, on the basis of equity and in accordance with their common but differentiated responsibilities and respective capabilities. Accordingly, the developed country parties should take the lead in combating climate change and the adverse effects thereof." This would require binding and real commitments leading to actual reductions in emissions – commitments that have not been made so far.

Still, the Climate Conference of Marrakesh removed the obstacles to the Kyoto Protocol coming into force at all. On March 4, 2002, the EU environment ministers decided to ratify the Protocol.

---

[20] *F.A.Z.* of 07-24-2001, no. 169, p. 1 f.; *Handelsblatt* of 07-23-2001.

4.1.2
The legal framework in European law

Treaties

In European law, sustainable development is touched upon as early as in the preamble to the EU and then in art. 2 indent 1 EU. It is called for, especially, in the basic provision of art. 2 EC in the version of the Amsterdam amendment to the treaty. This directive prescribes a sustainable development in the context of economic life and thus clarifies that such a development must not be guided one-sidedly by ecological considerations.[21] However, according to the integration clause of art. 6 EC "with a view to promoting sustainable development", ecological concerns must be considered consistently in connection with the policies referred to in art. 3 EC. The very dictate of integration formulated in that clause – which is, of course, limited to the requirements of environmental protection and which can be understood, at the same time, as an expression of the integrative approach immanent to sustainability – is suitable for implementing the obligation, addressed at all signatory parties, "to take climate considerations into account, as far as they are feasible, in their relevant social, economic and environmental policies ...", which is contained in art. 4 par. 1 lit. f UNFCCC. Since climate protection is among the "environmental protection requirements" cited in art. 6 EC, climate protection measures must be taken into account when laying down and implementing measures in the energy sector according to art. 3 par. 1 (u) EC[22]. Nevertheless, the postulate of sustainability goes further insofar as it derives its specific character from an abstract, equal integration of ecological considerations with economic and social ones. Hence, ecological concerns must not be considered with preference, neither exclusively nor a priori.[23] Consequently, a development is sustainable not just by being compatible with the environment and climate, and by preserving the resource base; beyond that, the social and economic components of the concept of sustainability demand the long-term security and stability of energy supplies (security of procurement) and the preservation of a competitive energy economy (competitiveness).[24] The objectives and standards of a sustainable energy policy, against which activities and instruments in the energy sector are to be measured, emerge from a dynamic balance between these determinants.

---

[21] More details in Frenz/Unnerstall, *Nachhaltige Entwicklung im Europarecht*, 1999, p. 155 ff., 177. Also see Badura, *Umweltschutz und Energiepolitik*, in: Rengeling (ed.), *Handbuch zum europäischen und deutschen Umweltrecht*, vol. II, 1998, § 83 Rn. 27.

[22] Still, a seperate energy policy does not exist, although it is possible for the purpose of climate protection based on the environment, as becomes clear from art. 175 par. 2 s. 1 indent 3; Steinberg/Britz, *Die Energiepolitik im Spannungsfeld nationaler und europäischer Regelungskompetenzen*, DÖV 1993, 313/314.

[23] More details in Frenz, *Sustainable Development durch Raumplanung*, 2000, p. 41 ff.

[24] See the Commission White Paper, "An Energy Policy for the European Union" of 1995-12-13, KOM (95) 682 fin.; Report of the German Federal Government to the UN Special General Assembly on Environment and Development, New York, 1997 – *Auf dem Weg zu einer nachhaltigen Entwicklung in Deutschland*, BT-Drucks. 13/7054, p. 38; Badura, *Umweltschutz und Energiepolitik*, in: Rengeling (ed.), *Handbuch zum europäischen und deutschen Umweltrecht*, vol. II, 1998, § 83 Rn. 1, 30, 34.

These statements of principle are supplemented in the section on environmental policy, most notably by the principles of precaution and prevention[25] according to art. 174 par. 2 s. 2 EC, which implies future-related action not least on the basis of probabilities supported by facts, as the principle of sustainable development also demands; Principle 15 of the Rio Declaration postulates action even if the factual basis is uncertain.[26] Referring specifically to resource economy, art. 174 par. 1 indent 3 EC stipulates a prudent and rational utilization of natural resources. Taking the needs of future generations into account is inherent to this demand.[27] However, it is still assumed that natural resources will be, and are, exploited.[28]

According to art. 174 par. 2 s. 2 EC, the direction of activities to be implemented according to joint decree or prescription by the member states must relate to the polluters, i.e. to the parties that, by their conduct, cause damage to the environment.[29] For this, factual evidence has to be available, even if some causal connections are still uncertain.[30] Also due to the polluter-pays principle, payments of subsidies to polluters require special justification.[31]

Programs for action

Particularly in the context of energy saving and the stabilization and reduction of CO_2 emissions, the concept of sustainable development has been taken up repeatedly in various documents by European bodies. Among them, the 5th Environmental Action Programme (EAP),[32] which regards changing existing attitudes as particularly important, deserves special emphasis. Considering the interaction between economic and social development, on the one hand, and the limited load capacity of the environment, on the other, it also reminds us of the necessity to strike a balance between human activity, development and environmental protection. The 5th EAP (1992–1999) was replaced, supplemented and developed further by the 6th EAP (2001-2010),[33] whose strategic emphasis is on describing the environmental goals and priorities of the EU strategy for a sustainable develop-

---

[25] On the synonymous content, see e.g. Kahl, *Umweltprinzip und Gemeinschaftsrecht*, 1993, p. 21 f.; Grabitz/Nettesheim, in: Grabitz/Hilf, EU, status: July 2000, art. 130 r Rn. 67; Rengeling, *Umweltvorsorge und ihre Grenzen im EWG-Recht*, 1989, p. 11; on the differences between the two terms, see e.g. Epiney, *Umweltrecht in der Europäischen Union*, 1997, p. 99.

[26] For more details, see Calliess, *Die neue Querschnittsklausel des Art. 6 ex 3 c EGV als Instrument zur Umsetzung des Grundsatzes der nachhaltigen Entwicklung*, DVBl. 1998, 559/563; Frenz, *Deutsche Umweltgesetzgebung und Sustainable Development*, ZG 1999, 143/146 ff.

[27] See, in the French version, *"prudente et rationelle"* and, in the Danish, "forsigtigt og fornunftigt".

[28] Treated exhaustively in Frenz, *Sustainable Development durch Raumplanung*, 2000, p. 32 ff.

[29] For more details on the extent of this principle, see Frenz, *Europäisches Umweltrecht*, 1997, p. 55 ff., w.f.N.

[30] See Di Fabio, in: *FS für Ritter*, 1997, p. 807/820 ff; Frenz, *Das Verursacherprinzip im Öffentlichen Recht*, 1997, p. 298 f.

[31] See under 6.1.

[32] *"Für eine dauerhafte und umweltgerechte Entwicklung. Ein Programm der Europäischen Gemeinschaft für Umweltpolitik und Maßnahmen im Hinblick auf eine dauerhafte und umweltgerechte Entwicklung"*, ABl. EG 1993, C 138, p. 1 ff.

[33] Communication by the Commission to the Council, the European Parliament, the Economic and Social Committee and the Committee of the Regions, on the 6th EAP, *"Umwelt 2010: Unsere Zukunft liegt in unserer Hand"* – Ein Aktionsprogramm für die Umwelt in Europa zu Beginn des 21. Jahrhunderts, 2001-01-24, KOM (2001) 31 fin.

ment.[34] Apart from economic prosperity (economic component) and a balanced social development (social component), the achievement of environmental policy goals, especially the prudent treatment of the natural resources of the earth and the protection of the global ecosystem, is a precondition for sustainable development.[35] This environmental policy dimension or component of the sustainability postulate is paid prominent attention in the 6th EAP, which defines the most important goals and measures of EU environmental policy within the time horizon of 2001-2010, with the highest priority on the accelerating global warming, caused by the greenhouse effect, and the resulting climate change. Hence, in contrast to the 5th EAP, whose climate protection strategy was still based on the stabilization of CO_2 emissions on the level of 1990, the 6th EAP designates as a long-term objective not only to ratify and implement the Kyoto Protocol (Union-wide reduction of greenhouse gas emissions by 8 % compared to the numbers for 1990, over the period leading up to 2008–2012) but, beyond that, to reduce the emission of greenhouse gases, especially CO_2, by 20 %–40 % of the 1990-values by 2020.[36] This is intended to take us closer, in the sense of a sustainable development, to the overall objective of stabilizing the concentration of greenhouse gases in the atmosphere at a level that excludes future unnatural fluctuations of the world climate (which requires a reduction by about 70 % compared to the 1990-figures[37]). More specifically, an EU-wide trade regulation on CO_2 emissions shall be developed by 2005; a further switch of electricity production from coal and mineral oil to other sources, natural gas in particular, involving lesser amounts of CO_2 emissions shall be achieved; and an increase in the contribution of renewable energies to electric power production shall be promoted.[38] This end is served, most notably, by the directive on the promotion of electricity from renewable energy sources in the internal electricity market, passed by the European Parliament and Council on 2001-9-7,[39] which lays down two general, direction-setting targets to be met by 2010: The share of renewable energies in the energy consumed in Europe must reach 12 %, where 22.1 % of the electricity consumed must come from regenerative energy sources. Furthermore, a system of certificates of origin for electric power from renewable energy sources shall be put into place. Electricity production from energy-saving combined heat and power systems is to be promoted, on the basis of the EU electricity directive,[40] in each member state. The predominant CO_2 emissions are to be subjected to a Union-wide system of trading in certificates.[41]

---

[34] KOM (2001) 31 fin., p. 3.
[35] See also KOM (2001) 31 fin., p. 11.
[36] KOM (2001) 31 fin., p. 28; as stated earlier by the EU Commisioner for the Environment, Ms.Wallström, in *FAZ* of 2001-01-25, no. 21, p. 7, "*Ehrgeizige Ziele beim Umweltschutz*".
[37] KOM (2001) 31 fin., p. 28.
[38] KOM (2001) 31 fin., p. 30.
[39] Passed by the EP on 2001-07-04; see also the Commission proposal for a directive, 05-10-2000 (ABl. 2000, C 311 E, 320).
[40] Directive 96/92/EC of the European Parliament and Commission of 1996-12-19, concerning common rules for the internal electricity market; compare, in particular, art. 8 par. 3 and art. 11 par. 3 therein.
[41] Proposal for a directive of the European Parliament and of the Council establishing a scheme for greenhouse gas emission allowance trading within the Community and amending Council Directive 96/61/EC, KOM (2001) 581 fin.

The European Union strategy for sustainable development

The Helsinki European Council of December 1999 invited the European Commission to prepare a "proposal for a long-term strategy dovetailing policies for economically, socially and ecologically sustainable development to be presented to the [Gothenburg] European Council in June 2001". After the Lisbon European Council (March 23/24, 2000) had issued the strategic goal for the Union "to become the most competitive and dynamic knowledge-based economy in the world, capable of sustainable economic growth with more and better jobs and greater social cohesion", the Stockholm European Council (March 23/24, 2001) decided that the environmental aspect should be added to the EU strategy for a sustainable development set out in Lisbon – a view that had already been expressed in the 6th Environmental Action Programme. The strategy for a sustainable development is to serve as a catalyst for political decision-making and public opinion in the coming years; it shall become the driving force towards institutional reforms and a far-reaching change in political, economic and public thinking, leaving behind the short-term, profit-oriented perspective and moving towards a longer-term mode of thinking and acting, taking account of the social and ecological consequences for the habitat and environment of human beings. The EU strategy for sustainable development[42], which was expressly welcomed by the European Council of Gothenburg (June 15/16, 2001)[43], thus constitutes the first comprehensive concept for sustainability on the European level. With it, the EU also fulfills a commitment given at the Rio Conference, according to which it was to develop strategies for sustainable development in time for the World Summit on Sustainable Development (Rio + 10), which was held in Johannesburg from September 2-11, 2002. Accordingly, the postulate of sustainability must become a core element in all fields of policy[44]. The strategy also lists the goals, priorities and (immediate) measures needed to achieve a comprehensive sustainable development, including those of importance for the energy sector and others intended to counteract climate change and increase the utilization of clean energies[45], in essence identical to those cited in the 6th EAP. On this list are, most notably, the gradual removal of subsidies for the production and use of fossil fuels (by 2010), a new framework for the taxation of energy, the introduction of the trade in CO_2 allowances, the promotion of alternative fuels, such as bio-fuels for cars and trucks, and measures to promote energy efficiency[46]. All dimensions of sustainable development according to the present EU strategy were to be revisited at each spring session of the European Council, the first of which took place in Barcelona, in spring 2002, and then revised comprehensively at the beginning of each new Commission's term of office. At the same time, as a supporting measure, the European Parliament and the Commission shall establish a committee and "round table", respectively, for sustainable development, the latter consisting of about ten independent experts representing a wide range of views and not bound by national or other political interests.

---

[42] Communication by the Commission: Sustainable development in Europe for a better world: the European Union's startegy for sustainable development, May 15, 2001, KOM (2001) 264 fin.
[43] Presidency conclusions from the European Council (Gothenburg), June 15/16, 2001, p. 4.
[44] KOM (2001) 264 fin., p. 7.
[45] KOM (2001) 264 fin., p. 12.
[46] For details, see KOM (2001) 264 fin., p. 12 f.

4.1.3
Constitutional framework

In the German constitution (*Grundgesetz*, GG), the concept of sustainability in specific areas follows, mainly, from art. 20 a, where "protecting the natural fundamentals of life" is laid down also with regard to the responsibility for future generations.[47] For the present generations' long-term responsibility, which is explicitly laid down in the state definition of environmental goals, links this definition to the central postulate of sustainability, which is, according to the original, still formative definition by the Brundtland Commission: [to strive for] "a development that meets the needs of the present without compromising the ability of future generations to meet their own needs". Thus, the climate[48] is protected even if there is no individual legal reference and, hence, no compelling duty of protection on the part of the state through the legal concept of constitutional duties of protection derived from art. 2 par. 2 in connection with art. 1 GG.[49] The generations living today have to curb their actions so that later generations will have the fundamentals of life at their disposal, which would enable them to prosper. Of course, this does not translate into a concrete measure of the economizing and safe-guarding required. It is up to the legislators to concretize this component of the state definition of environmental goals for specific areas, always taking into account other constitutional stipulations. For the energy sector, e.g. energy saving measures may be considered.

The economic and social components of the principle of sustainable development are, as such, not included in art. 20 a GG, but can be gathered from art. 12, 14 GG and art. 20 par. 1 GG, respectively. Thus, the ecological side is balanced by two counterpoints of equal weight, which must be taken into account as of equal importance in a comprehensive weighing process, in order to avoid a bias towards environmental issues.[50]

4.1.4
A duty to protect the environment?

The individual elements of art. 174 par. 1 ECT describe the contents of environmental policy in more detail and direct this policy towards certain goals. Considering the upward revaluation of environmental protection in art. 2, 3 ECT as well as in art. 130 s ECT, as far back as with the Maastricht Treaty, the phrasing "community policy on the environment shall contribute to the pursuit of the following objectives" would not have implied a relativization compared to the precursor provision of art. 174 par. 1 ECT.[51] Hence, the environmental policy assumed as such is bound

---

[47] See Kloepfer, in: *Bonner Kommentar*, status: 77. Lfg. 1996, art. 20 a Rn. 58 ff.; Epiney, in: v. Mangoldt/Klein/Starck, GG II, 4. edition 2000, art. 20a, Rn. 30, 97.

[48] Scholz, in: Maunz/Dürig, status: 35. Lfg. 1999, art. 20 a Rn. 36; Epiney, in: v. Mangoldt/Klein/Starck, GG II, 4. edt. 2000, art. 20 a Rn. 18; Murswiek, NVwZ 1996, 222/225.

[49] Steinberg, *Verfassungsrechtlicher Umweltschutz durch Grundrechte und Staatszielbestimmungen*, NJW 1996, 1985/1991; but also see Kruis, *Der gesetzliche Ausstieg aus der „Atomwirtschaft" und das Gemeinwohl*, DVBl. 2000, 441/443.

[50] Frenz, *Nachhaltige Entwicklung nach dem Grundgesetz*, in: Hendler/Marburger/Reinhardt/Schröder, *Jahrbuch des Umwelt- und Technikrechts* 1999, UTR 49, 1999, p. 37, 50 ff.

[51] Epiney/Furrer, EuR 1992, 369 (381 f.).

by the four elements cited there. If these are not in place, the Community needs to take action.[52] Still, the objectives will never be achieved entirely. From that perspective, permanent action by the community is required. For this reason, and due to limited organizational resources, it is hardly possible to work towards the goals listed in art. 174 par. 1 ECT in every area. Hence, the Community is not obliged to act in specific cases,[53] and the "market citizens" have no right to Community action and, in turn, are not entitled to environmental protection; art. 174 ECT cannot be applied directly.[54]

Art. 20 a GG also describes a target quantity for sustainable development, especially for its intergenerational component; it is intended for implementation through the legislator, who, as is the case on the Community level, faces the problem that he cannot take action in all areas at the same time. The constitutional protection commitments relate more closely to the individual, to protect human health[55] as well as property[56]. They demand environmental protection measures, also in the interest of later generations,[57] but only as such, not in any concrete manifestation; the more specific shaping of them is a matter for the legislator. Thus, these commitments demand the maintenance of a minimum standard, an "ecological subsistence level".[58] However, they give grounds for subjective claims only if the due measures do not materialize, or if they are clearly insufficient for meeting, or even coming close to, the protection target.[59] Still, regarding sustainable development, this has to be welcomed even if the measures do not promise a lasting effect and comprehensive environmental protection without shifting the pollution from one environmental medium to another.[60]

4.1.5
Implementation in energy law

Foundations for a sustainable development also appear in more recent legislation concerning energy law in Germany. For instance, the purpose of the Energy Economy Act (*Energiewirtschaftsgesetz*, EnWG)[61] is to provide electricity and gas as safely, cheaply and environmentally sound as possible, in the common interest. This touches upon economic aspects, but also refers clearly to environmental issues, for

---

[52] Also for instance Grabitz/Nettesheim, in: Grabitz/Hilf, art. 130 r Rn. 10.
[53] Anders Epiney, *Umweltrecht*, p. 95 f.; see also Heinz/Körte, JA 1991, 41 (43); Kahl, *Umweltprinzip und Gemeinschaftsrecht*, p. 94; Lietzmann, in: Rengeling, *Europäisches Umweltrecht und europäische Umweltpolitik*, p. 163 (174 f.).
[54] Grabitz/Nettesheim, in: Grabitz/Hilf, art. 130 r Rn. 13; also see Krämer, EuGRZ 1998, 285 (291).
[55] Already BVerfGE 53, 30 (57); 56, 54 (73); BVerfG, NJW 1996, 651.
[56] BVerfG, NJW 1998, 3264 (3265).
[57] Frenz, *Nachhaltige Entwicklung nach dem Grundgesetz*, in: Hendler/Marburger/Reinhardt/Schröder, JUTR 1999, S. 37, 65 f.
[58] Kloepfer, *Umweltrecht*, 2. edition. 1998, § 3 Rn. 38.
[59] BVerfGE 77, 170 (215); 92, 26 (46); BVerfG, NJW 1996, 651.
[60] More in Frenz/Unnerstall, *Nachhaltige Entwicklung im Europarecht*, 1999, p. 202.
[61] Act concerning the supply of electricity and gas *(Energiewirtschaftsgesetz – EnWG)* of 1998-4-24, BGBl. I p. 730, especially as amended by art. 2 of the act granting priority to renewable energy sources *(Erneuerbare-Energien-Gesetz – EEG)* and in amendment of EnWG and the act covering mineral-oil taxes of 2000-3-29, BGBl. I p. 305.

environmental soundness means, according to § 2 par. 4 EnWG, that energy provision meets the requirements of dealing with energy in a rational and economical manner, ensuring the sparing and sustainable use of resources, and putting as little strain as possible on the environment. The proviso of dealing with energy in a rational and economical manner runs parallel to art. 174 par. 1 indent 3 EC, while the stipulation of a sparing and sustainable use of resources concretizes the protection of the natural fundamentals of life also in responsibility for future generations, as demanded in art. 20a GG. The same applies to the requirement to put as little strain as possible on the environment. These general stipulations must be fulfilled by the basic decision in the Energy Economy Act for a far-reaching liberalization of the electricity market and for restricting the state, essentially, to the role of a guarantor.

The particular importance of using combined heat and power systems and renewable energies may be emphasized in § 2 par. 4 s. 2 EnWG, but, according to § 2 par. 5 EnWG, the commitment to purchase electricity from these and other renewable energies, and to feed it into the general supply grid, is left to another, dedicated act of parliament. The act granting priority to renewable energy sources[62] (§ 1) states as its purpose the facilitation of a sustainable development of energy provision in the interest of climate and environmental protection. Here, the concept of sustainability is referred to explicitly. The intended instrument is to increase significantly the share of renewable energies in the provision of electricity so that the share of such energies in the total energy consumption is at least doubled by 2010, in compliance with the goals set for the European Union and Germany. To achieve this, the grid operators are put under obligation to purchase preferentially all the electric power offered from installations producing renewable energies, and to pay for it at fixed minimum rates significantly above the usual rates for conventional energies. The obligation covers hydroelectricity, wind power, solar radiation energy, geothermal energy, waste dump gases, sludge gas, coal mine gas and biomass. This supportive regulation entails various issues concerning European law with regard to the free movement of goods and the prohibition of state aid.[63]

4.1.6
Implementation in regional planning and mining law

Regulating the exploitation of natural resources is another central element of a sustainable energy policy. Such regulation determines the resource stocks that will be left for future generations. Statutory stipulations in this respect follow from regional planning and mining law. The Regional Planning Act (*Raumordnungsgesetz,* ROG)[64], especially, which also regulates the planning of locations where raw materials for the production of energy are exploited or where centers for regenerative energy production are to be established, is a particularly successful example of implementing the concept of sustainability. According to § 2 par. 2, it is subject to

---

[62] Renewable energies act (*Erneuerbare-Energien-Gesetz* – EEG) promulgated as art. 1 of the aforementioned act of 2000-3-29, BGBl. I p. 305.
[63] See also chapter 6.3.
[64] Of 1997-8-18, BGBl. I p. 2081.

the guiding concept of sustainable regional development, which reconciles the social and economic demands on land with its ecological functions and leads to a sustained order that is balanced over large areas. Hence, sustainable development, in terms of the Regional Planning Act, aims at balancing the ecological, economic and social functions of the land with the demands on the land. This means none of the three aspects per se is treated with preference a priori. The legislator thus rejected the demand that systems should be optimized in favor of one aspect. This guiding concept acts as the principal rule of interpretation and application. Accordingly, relevant and diverging interests (principles) have to be reconciled in keeping with the guiding concept. The eight component aspects detailed in § 1 par. 2 s. 2 ROG are meant to clarify this central guiding concept.[65]

The aspect cited under no. 1 demands that regional planning "ensures the right to personal self-fulfillment ... with responsibility for future generations". Accordingly, not only the present effects and risks for the functions of the entire territory of Germany and her regions must be considered, but also, and especially, the long-term effects and risks must be appreciated and taken into account when considering planning decisions. This becomes particularly clear in the case of exploiting mineral resources for energy production. The present consumption of non-regenerative resources reduces, or even excludes, possibilities for later generations to use the same resources. Thus, due to its long-term consequences, the use of non-renewable resources is fraught with questions especially with regard to long-term responsibility. For case constellations like this, in particular, the first component aspect of § 1 par. 2 s. 2 ROG gives grounds for the legal obligation of planners to take the interests of future generations into account when making planning decisions. Thus, according to this aspect, sustainable land development is to guarantee future land demands and land functions for generations to come. Conversely, the unhindered self-fulfillment of the present population, which includes the right to exploit resources, is assumed in § 1 par. 2 s. 2 no. 1 ROG. The issue is, hence, the limitation, not the total suppression of the extraction of natural resources, even if they are non-renewable. The extent of the necessary self-restriction derives from the circumstances of each individual case, which must be taken into account when making a considered planning decision. It cannot be laid down as a static entity, but must be determined dynamically, depending on the deposits presently available, the deposits that can be developed in the future, and the expected needs of present and future generations.[66]

The second aspect of § 1 par. 2 s. 2 ROG authorizes and obliges the state planning agencies to protect and develop the natural fundamentals of life; it thereby emphasizes the ecological dimension of the principle of sustainable regional planning. The assumption under § 1 par. 2 s. 2 no. 3 ROG – that the locational prerequisites for economic developments are created – makes clear that the legislatorial conception of a mandate to exercise precaution, as applied in regional planning law, cannot be interpreted as plainly prohibiting any degradation. The purpose is, rather, to reconcile the social, economic and ecological demands on the land, which should

---

[65] According to the reasons given for the draft bill tabled by the German federal government, BR-Drucks. 635/96, p. 40.

[66] More in Frenz, *Sustainable Development durch Raumplanung*, 2000, p. 152 ff.

lead to a "sustained order that is balanced over large areas". Another aspect to clarify the principle of sustainable development is the demand to create the locational prerequisites for economic developments. This mandate reflects the economic demands that humans place on the land.[67] Thus, according to this legislatorial conception, the [preservation] of the natural, and the development of the "economic", fundamentals of life are planning guidelines of equal rank. The fourth guideline stipulates that the "land use possibilities shall be kept open in the long term". This guiding concept is designed to take into account the necessity of a long-term precautionary policy concerning land use.

The concept of sustainability is also discernible in the Federal Mining Act (*Bundesberggesetz*, BBergG), which is the second piece of legislation relevant to the regulation of resource extraction. It regulates the exploitation of energy resources, in particular, such as coal and gas in Germany. Still, it does not refer openly to sustainability. Being a traditional field of legislation, mining law seems to be tailored exclusively to the needs of resource exploitation. However, this view is shortsighted. Ensuring the security of procurement for today is not the only standard to be met. Elements of sustainability already shine through in the purpose provision of § 1 BBergG. An inspection of the provisions regulating the extraction of natural resources also shows that these can be interpreted in the spirit of sustainable development.[68] The purpose provision of § 1 BBergG, which acts as an interpretation rule, aims at a long-term, precautionary protection of resource deposits and takes account of a view guided by the intergenerational component of the concept of sustainability with regard to mining issues. The medium- to long-term security of natural resource procurement thus intended primarily comprises an economic and social component. The economical and sparing treatment of land and soil, demanded by § 1 no. 1 BBergG, takes up a point that is commonly regarded as a secondary aspect of the sustainability concept, which serves, above all, as a safeguard against the erosion of ecological interests.

The present mining law also contains a number of "opening clauses", through which issues external to mining law, especially those from environmental and planning law, flow into the approval and licensing process. The reference to "public interests" contrary to mining interests in numerous provisions of mining law requires a reconstructive balancing decision by the mining authority. In this way, the duty to manage conflicts through weighing conflicting interests against each other – which is inherent to sustainability – can be honored.

4.1.7
International obligations concerning energy security

Presently, about 50% of the energy demands of the European Union are covered by imports; this corresponds to 6% of the total import volume. In geopolitical terms, ca. 45% of the oil imports originate from the Middle East; 40% of the natural gas

---

[67] According to the brief reasoning for the bill presented by the federal government, BT-Drucks. 13/6392, p. 79.
[68] Details, also on the following, in Frenz, *Bergrecht und Nachhaltige Entwicklung,* 2001, p. 11 ff.

imports stem from Russia. By 2020 – according to a EU prognosis that assumes status-quo conditions – the import component will have risen (again) to 70% (KOM (2000) 769 fin.). Even more dramatic is the shift to renewed dependence on the Middle East, where about two thirds of future oil reserves are located (see fig. 4.4 below) and where, as estimated by the International Energy Agency (IEA), more than 85% of any additional production capacity is likely to be found. The reasons for this development are the extremely high costs of developing conventional and alternative oil resources outside OPEC (e.g. tar sands in Canada) and the still growing demand for oil. The latter, in turn, is caused by the fact that, by now, oil is used predominantly in an energy market growing particularly fast, i.e. the transport sector.

The realization that mobility – a basic prerequisite of a work sharing, global economy – depends on the politically unstable Middle East may give rise to unease, but not to action (yet) to slow down the growth in the demand for oil in the transport sector. Most notably in the US, where gasoline consumption per 100 km is 2–3 liters higher than in Europe, hardly any effort appears to be made to reduce consumption, although energy imports will also continue to rise there – regardless of occasional attempts to increase energy autarky within NAFTA (see fig. 4.5 below).

Only the International Energy Agency (IEA), founded in the aftermath of the first oil crisis of 1973/74, is concerned specifically with energy security, or the security of oil procurement, in particular. The original task of this international organization with, presently, 26 member states, is to develop joint measures to prevent bottlenecks in the supply of mineral oil. Accordingly, the member states have committed themselves to keep oil reserves corresponding to at least the net imports over 90 days of the preceding year. In addition to this, further rapid and flexible emergency measures (e.g. switching to other energy carriers, and programs for restricting demand and boosting production) are set out in the Coordinated Emergency Response Measures (CERM).

In 1993, the ministers of the member states adopted the "Shared Goals", in which the guiding principles are laid down: security of procurement, environmental protection and economic growth. Long-term security of procurement shall be ensured by

– diversifying the energy systems,
– increasing the efficiency of all types of energy,
– flexibility,
– restricting energy consumption
– and promoting renewable energy carriers.

The promotion of non-fossil energy sources is afforded a high priority. Research, development and market penetration as well as the transfer of new and improved technologies are regarded as necessary elements for achieving the goals. Apart from new, flexible solutions, reforms of regulation regimes are also to contribute. According to the IEA, the deregulation of the energy sector in the member states has already resulted in improved efficiencies and opened up new opportunities for innovation (IEA 2001b).

The objectives of the EU concerning the energy sector – global competitiveness, security of procurement and environmental protection – are largely identical to those of the IEA. A debate began in the year 2000, when the Green Paper, "Towards a European Strategy for the Security of Energy Supply" (KOM (2000) 769 fin.) was published. Presently, the recommendations are under discussion. Binding provisions have not been formulated yet, but the following proposals were submitted:

- Taxation as an instrument to control demand
- Energy savings and diversification in the construction and transport sectors
- Expansion of new and renewable energy carriers (Support through grants, tax reliefs and financial aids)
- New import routes for mineral oil and gas, as well as increasing the reserves

4.2
Evaluation of the global energy system under criteria of sustainability

All forms of sustainability imply handing over a certain potential for use to succeeding generations (chapter 2). The various definitions of sustainability (very weak to very strong sustainability) only differ in their interpretation of this potential. In the following, we assess the sustainability of the energy system by analyzing the present energy system in the context of different scenarios for the future, using the concept of critical sustainability, as defined in chapter 2, as a standard of reference.

4.2.1
Characteristics of the present energy system

Through the last 50 years, commercial energy consumption has grown by a factor of 5 worldwide, to about 350 EJ[69] per year (fig. A.1 in appendix 1), to which the non-commercial use of energy (firewood, biological waste etc.) has to be added. The World Resources Institute (WRI) cites an annual total (commercial and non-commercial) energy use of 380 EJ for the year 1997. In terms of continuous power, this corresponds to a global consumption of 12 Terawatt (TW)[70] or ca. 2,100 watts per capita (fig. A.2, appendix 1).

Today's global energy system is based mainly on the non-renewable energy resources coal, mineral oil and natural gas, whose total share amounts to 83 % (95%) of the total (commercial) energy consumption (fig. A.2, appendix 1). Nuclear and hydroelectric power contribute 2.3% and 2.4%, respectively, to meeting the demand.[71] New, regenerative energy sources such as solar energy, wind energy, geothermal energy and others, represent a marginal contribution of 0.4% in total. The largest non-fossil contribution to the energy system is provided by biomass, which stems mainly from non-commercial energy resources. The latter play an important, quite often the most important role especially in developing countries, even though they cause other, negative ecological side effects in many cases (e.g. deforestation and desertification).

---

[69] EJ = 1018 joules. See chapter 2 (Box 1) for the definitions of energy units.
[70] Terawatt = 1012 watts; 1 watt = 1 joule per second.
[71] Hydroelectric and nuclear power are quantified here by means of the electricity produced, without the multiplicator 3 (the reciprocal of the average efficiency of thermal electricity production), which is used in some energy statistics.

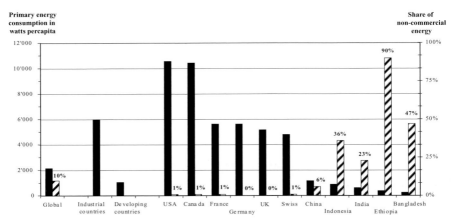

Fig. 4.1 Primary energy consumption, global and for selected regions and countries. Black bars: Primary energy consumption in Watt per capita in 1997. Hatched bars: Share of non-commercial energy in percent of the primary energy consumed in 1993 (global: 1997). Data sources: WRI (1997), WRI (2001), BP (2001).

Figure 4.1 confers an impression of the large global differentials in energy use between the poorest developing countries, e.g. Bangladesh (260 Watt/capita), and the richest industrial nations such as USA and Canada (more then 10,000 Watt/capita). The energy consumption per capita diverges by more than a factor of 40. Furthermore, when taking into account commercial energy only, this difference increases by at least another factor of 10.

In the EU, the mean energy consumption per person is about 4,500 Watt (fig. B.1, appendix 1). In Germany, the primary energy consumption amounted to 5,100 Watt per person in 1997, of which the fossil energy carriers provide a share of 93.6%, compared to 4.6% coming from nuclear energy. With a share of less than 2%, renewable energies only make a marginal contribution to the primary energy consumed (fig. B.2, appendix 1).[72]

4.2.2
Prognoses for the development of the global energy system over the coming 100 years

Various organizations and research institutes have produced forecasts for the development of energy demand up to the year 2100. Every single one predicts that the world population and the global gross domestic product (GDP) as well as the GDP per capita and the energy consumption per capita will grow significantly. The prognoses differ, predominantly, in their assumptions concerning the efficiency gains made in the use of energy, i.e. the decoupling of the rise in energy consumption

---

[72] *Umweltbundesamt* (2001): In the list shown there, the share of nuclear energy is three times larger, since it is expressed as the overall heat produced by nuclear power plants. This value is roughly three times the electricity production.[4]

from economic and population growth, and with respect to the role of the so-called renewable energies (mainly solar energy in all its variations) as well as regarding population growth. A selection of such models is presented in section C of appendix 1 (box 1 and 2, tab. C.1), where six scenarios, developed by the International Institute for Applied Systems Analysis and the World Energy Council (IIASA/WEC scenarios) and representing the typical variation width of such models, are compared (app. 1, fig. C.1 to C.7). The analysis that follows is based on the IIASA/WEC B scenario, which is in the middle portion of the distribution; it assumes average economic growth and technological development and does not take into account any explicit climate protection measures.

Apart from decoupling economic growth from energy consumption, *decarbonization* (decoupling of CO_2 emissions from energy consumption) - through replacing fossil energy sources with renewable and nuclear energies or by sequestering carbon dioxide in the oceans or in the geological subsoil - plays a central role. Focusing on CO_2 in the energy sector is reasonable because carbon dioxide amounts

---

Box 4.1: Relationships between the rates of change in global energy demand, global atmospheric CO_2 emissions and CO_2 intensity

A. The total, global energy demand per year (E) can be expressed as:

$$E = \left(\frac{E}{GDP}\right)\left(\frac{GDP}{N}\right)N = \left(\frac{GDP}{E}\right)^{-1}\left(\frac{GDP}{N}\right)N = \frac{q_w N}{\varepsilon} \qquad (1)$$

with the following definitions:

$\varepsilon = \dfrac{GDP}{E}$ Energy efficiency (GDP per unit of energy consumed)

$q_w = \dfrac{GDP}{N}$ Income ratio (GDP per capita of the population)

N Population

The relative rates of change of the quantities in eq. 1 are related through the following equation:

$$k_E = k_N + k_q - k_\varepsilon \qquad (2)$$

with the relative growth rates:

$k_E = \dfrac{1}{E}\dfrac{dE}{dt}$ relative growth of the total energy demand (yr^{-1})

$k_N = \dfrac{1}{N}\dfrac{dN}{dt}$ relative population growth (yr^{-1})

$k_q = \dfrac{1}{q_w}\dfrac{dq_w}{dt}$ relative growth of income ratio (yr^{-1})

$k_\varepsilon = \dfrac{1}{\varepsilon}\dfrac{d\varepsilon}{dt}$ relative growth of energy efficiency (yr^{-1})

Conclusion from eq. 2: For k_E to become negative, the growth rate of the energy efficiency must exceed the sum of the growth rates of the population and the income ratio.

B. The total atmospheric CO_2 input per year (C) can be expressed as:

$$C = \left(\frac{C}{E}\right)\left(\frac{E}{BIP}\right)\left(\frac{BIP}{N}\right) N = \left(\frac{E}{C}\right)^{-1}\left(\frac{BIP}{E}\right)^{-1}\left(\frac{BIP}{N}\right) N = \frac{q_w N}{\gamma \, \varepsilon} \qquad (3)$$

$\gamma = \dfrac{E}{C}$ $\qquad\qquad$ CO_2 efficiency of the energy system (energy per CO_2 emission)

The relative rate of change in CO_2 emission C satisfies the relationship:

$$k_C = k_E - k_\gamma = k_N + k_q - k_\varepsilon - k_\gamma \qquad (4)$$

with $\quad k_\gamma = \dfrac{1}{\gamma}\dfrac{d\gamma}{dt}$ $\qquad$ **Decarbonization rate**

Conclusion from eq. 4: For carbon emissions into the atmosphere (C) to decrease, the decarbonization rate k_γ must exceed the growth rate of the total energy demand k_E.

Conclusion from eq. 4: For carbon emissions into the atmosphere (C) to decrease, the decarbonization rate k_γ must exceed the growth rate of the total energy demand k_E.

C. CO_2 intensity is defined as CO_2 emissions per GDP, i.e.

$$CO_2 \text{ intensity} = \frac{C}{BIP} = \frac{C}{E}\frac{E}{BIP} = \frac{1}{\gamma\varepsilon}$$

to 95% of the climate-relevant emissions accrued in that sector. In order to get a better quantitative understanding of the decoupling mechanisms, some simple mathematical relationships, which allow an analysis of the decoupling mechanisms being discussed, are summarized in box 4.1.

The IIASA/WEC B scenario is shown in fig. 4.2, as an example of a prognosis for the coming 100 years (see appendix 1, section C). According to this model, global energy demand would increase 3.5-fold between 2000 and 2100. In simple terms, this rise would be brought about by both the global population and the energy consumption per person just doubling. Simultaneously, the gross domestic product (GDP) per capita would grow by a factor of 4. This growth, combined with the energy demand per person roughly doubling, would mean that energy efficiency would also double. However, even in 2100 the discrepancies between the industrial nations and the developing countries would still be vast, both with regard to GDP per person (a factor of 4.8) and energy consumption per person (a factor of 2.3).

Since, according to fig. 4.2, population growth will diminish over the next 100 years, the formalism in box 4.1, which is based on exponential growth curves, is only applied to the period from 2000 to 2030. This approach yields the following specific rates of change:[73]

Population growth	k_N	0.012 yr^{-1}	(1.2% per year)
Income ration growth	k_q	0.009 yr^{-1}	(0.9% per year)
Energy efficiency	k_ε	0.007 yr^{-1}	(0.7% per year)
Growth of the total energy demand	k_E	0.014 yr^{-1}	(1.4% per year)

Hence, according to this scenario, a mere third of the growth in energy demand due to the growth of populations and incomes ($k_N + k_q = 0.021$ yr^{-1}) would be compensated by an increase in energy efficiency ($k_\varepsilon = 0.007$ yr^{-1}). To keep constant the atmospheric CO_2 input, while the total energy demand grows by 1.4% annually, the decarbonization rate, k_γ, would also have to be 1.4% yr^{-1}, (box 4.1, eq. 4). However, according to the prognoses of the climate models, global CO_2 emissions 100 years from now should be at only about 40% of today's rate (see tab. 4.6), which would necessitate an annual reduction of 0.9% of such emissions. To achieve this reduction target despite the predicted growth in energy consumption, the global decarbonization rate would thus have to be even higher, at 2.3% (1.4% + 0.9%) per year. Presently, the figure is only 0.3% per year (Grübler und Nakićenović 1997). In fact, considering the present imbalance between north and south, the annual reduction in CO_2 emissions in the EU and other industrial countries would have to reach the figure of 2%. Possibilities for decarbonization are discussed in more depth in chapter 5.

4.2.3
Excursus: Electricity, deregulation and sustainability

Electricity occupies a key position in all energy systems. It is an absolute prerequisite for the development of any country. In the industrial countries, it reflects the trend towards the service sector and the growing importance of private energy consumption as compared to the industrial demand. Electricity production amounts to 15% of the global consumption of commercial primary energy (tab. 4.1). 63.7% of electrical power is produced from fossil fuels. Due to the thermodynamic limits on the efficiency of thermal power plants, this means nearly 30% of the fossil energy resources used each year is burned to produce electricity. Hydroelectric and nuclear power provide a share of ca.17% each of the final electrical energy. The new, renewable energies (geothermal, wind, photovoltaic and others) contribute the remaining 1.6% to electricity production. In times to come, they would have to play a central role in a sustainable energy system. Yet, the substitution in the direction of renewable energy resources, which is getting underway only slowly, is still struggling with economic difficulties. This we will demonstrate, in an exemplary fashion, by looking at the effects of deregulation in the European electricity market.

---

[73]   For the shorter period (2000/2020), the growth coefficients would be very similar.

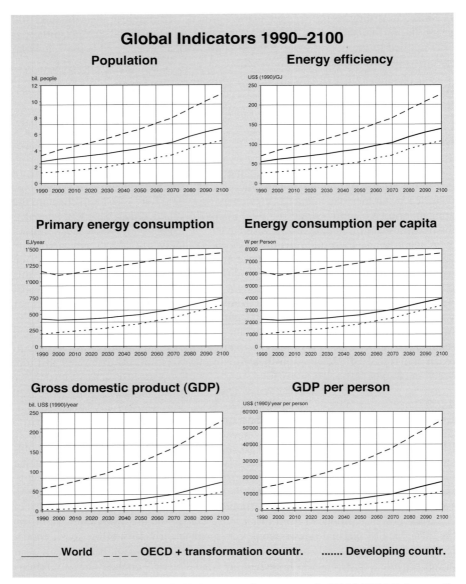

Fig. 4.2 Global indicators 1990–2100 (corresponding to scenario IIASA/WEC B in app. 1). Data source: Nakićenović et al. 1998

Table 4.1 The importance of electricity in the global energy system

Global consumption of commercial primary energy (1999)	**11.3 TW**	
Electricity production	**1.7 TW**	(15%)
Coal/Wood	*38.1 %*	
Mineral oil	8.5 %	
Natural gas	17.1 %	
Total fossil	*63.7 %*	
Nuclear energy	17.2 %	(290 GW)
Hydroelectric energy	17.5 %	
Others (geothermal, wind, photovoltaic etc.)	1.6 %	

Data source: IEA (2001a)

The liberalization of the European markets for grid-bound energies, which began in the 1990s, triggered serious restructuring in this sector. Grid-bound energy, as a "natural" monopoly, was beyond competition in almost every European country. The confined markets ("demarcation boundaries") were regulated through a multitude of rules. (Electricity prices, for instance, were endorsed on the basis of proven costs plus a profit margin).

Already in the run-up to liberalization, energy suppliers initiated considerable rationalization measures in order to be able to offer competitive prices. However, price wars and destructive competition following the opening-up of the market could not be foreseen in their actual fervor. In Germany, the electricity prices were at the level of the short-term marginal costs at times during the year 2000 (Haupt and Pfaffenberger 2001). Considerable overcapacities – about 40 to 50 GW in continental Europe – are a decisive factor in the price collapse. Previously, returns on investment in the energy sector had been secured through regional monopolies. Now, as a consequence of liberalization, decisions concerning the energy industry were characterized by uncertainties and more stringent demands on profitability, as is the case in all other markets that are based on competition. The capital costs, previously at 5% to 6 %, now amount to 9% to 12% depending on the conditions of financing, which changes the basis of investment calculations.

The electricity industry is very capital-intensive. The depreciation rate is roughly double of that in [other] manufacturing industries. On average, the costs amount to about a quarter of the prime cost of electrical power. They vary widely, depending on the type of installation: For nuclear power plants, the base load production incurs capital-dependent costs of ca. 50%. For hard-coal fired steam power plants, the figure is about 32%, and for combined gas and steam turbine plants it amounts to just about 12%. The number is about 90% for transmission grids. The significant shift in price ratios, especially concerning capital costs, and the competitive environment should lead to significant shifts in the electricity sector.

Effects on fuel use: Due to changed framework conditions, the economic advantages of small and efficient gas-fired plants have become more important (see e.g. Schlesinger and Schulz 2000, p. 109, and Lapidus et al. 2000, p. 635). In contrast to the high investment costs, long amortization periods and low operating costs of large power plants, gas and steam turbines are characterized by low investment costs and long amortization periods, but also by high fuel costs at about 75% to 80% of the prime costs for electricity. A further advantage is their flexibility of use: They can be switched into the grid whenever the load profile demands it. In an EU-commissioned study on the economic efficiency of various options and the effects of taxes and subsidies, eight typical electricity production technologies in 14 European countries were analyzed (Capros et al. 2000): a pressurized cycloid furnace plant representing "clean" coal-based technology, a monovalent lignite power plant, a power plant fueled by low-sulfur fuel oil, a monovalent thermal power plant fueled by biomass or waste, land-based wind turbines at windy locations, photovoltaic cells, and a pressurized water reactor. The results are evidence for the trend towards substituting gas for coal. At an annual duty cycle of 7,000 hours, the pressurized cycloid furnace plant (fueled by imported coal) shows the lowest production costs, compared to gas technology, in only five European countries. The lower the usage rate, the stronger the shift in favor of gas as the preferable fuel.

Hence, as natural gas involves considerably lower CO_2 emissions per unit of electrical energy produced than coal, liberalization triggers a trend towards decarbonization, since the advantages of a flexible power provision and the quick amortization of small, gas-fired plants are crucial with regard to new investments for peak and medium load production. For the production of electrical power in the base load domain, on the other hand, some European countries will probably continue to use large coal-fired plants that incur low fuel costs.

Effects for combined heat and power production: Combined heat and power plants (CHPs) offer ecological advantages since, by their efficient use of waste heat, they contribute to reduced energy consumption and, thus, to climate protection. The quantities crucial for their economic efficiency are the number of hours operated under full load and the process heat credit. Due to the slump in electricity prices, this technology has come under intense pressure now. In Germany, for example, CHP plants struggle for profitability as, generally, the proceeds needed for covering the costs cannot be realized. Economic efficiency is achieved only when a plant has been written off and offers adequate efficiency (Besch et al. 2000). The construction of new plants remains unlikely as long as electricity prices do not reflect the full costs. Presently, capacities for combined heat and power are reduced; industrial and communal plants, in particular, are shut down. The market for these types of power plants in Germany shrank by 50 % in 1999. This affected larger (central) CHPs, in particular, since the long-distance heating grid needed for their operation ties up large investment costs and suffers higher heat losses then decentralized CHPs. For the medium term, the latter should play an important role in a sustainable electricity supply system.

Effects on the use of regenerative energy sources: Electricity from regenerative sources also faces an opened market and competition and hence is equally challenged concerning its economic efficiency. Table 4.2 shows clearly that the regenerative energies are not competitive at present price levels. This situation is aggravated by the price slump induced by liberalization, which further widened the gap between fossil and renewable energy carriers. The best starting position, in relative

Table 4.2 Electricity prime costs in Eurocents per KWh in Germany

Non-regenerative	
Average costs	ca. 3.6
Wholesale price	ca. 2–2.6
CHP	3.7–4.8
Regenerative	
Wind	4–15
Biomass	6–10
Biogas	6–15
Geothermic	8–10
Decentralized hydroelectric...	6–13
Photovoltaic	60–90

Source: Besch et al. (2000), *Bundesumweltministerium* (2000)

terms, is enjoyed by wind generators at very favorable sites. At € 0.04 per KWh, their production costs are only double the present wholesale price for electricity. We have seen an enormous growth of capacities in this area in Germany, too. However, without subsidies or other instruments of environmental policy, the existing technologies (apart from large hydroelectric power plants that have been written off) for the environmentally sound production of electricity are not competitive.

Still, the future development of regenerative energy sources and the electricity market on the whole must not be judged from the cost perspective alone, since the liberalization of the electricity markets still does not offer the chance to engage potentials that did not exist previously. New actors have entered the market with novel products such as "green electricity" and challenge established companies to compete in this market segment. The established players try to present themselves as "good corporate citizens" under these new conditions. Many of them contribute to the development of regenerative energy sources. Even the term "energy services provider", which had been frowned upon in the past, was revived as a key strategic concept. Nevertheless, it remains doubtful whether the former monopolists will really be the "pathfinders" to a new energy system or if they will rather raise the convenience demands of (end) users by offering new products.

4.2.4
Assessing sustainability

A variety of sustainability concepts were discussed in chapter 2.1.3. The present considerations are based on the concept of critical sustainability, which assumes a limited substitutability of natural and material capital and starts from the concept of a critical natural capital marking the limits of substitutability. Thus, this concept does not exclude completely the degradation of natural capital (e.g. through consuming non-renewable energy resources) insofar as such consumption does not lead to falling below safe minimum standards, and is compensated by a growth in material and human capital. In the following, we will analyze the present energy system with regard to critical sustainability.

Climate changes

The atmospheric CO_2 concentration has risen by 31 % since 1750, from 280 ppm[74] to the present value of 367 ppm. This is more than the maximum during the last 420,000 years, which was reconstructed from paleontological data. Three quarters of the CO_2 emissions of the last twenty years were caused by the burning of fossil fuels. The remainder is a result of land-use changes, especially woodland clearing and intensive animal husbandry. The use of fossil energy carriers, as practiced today, is inherent to the rise in atmospheric CO_2 concentrations and thus linked to interference in the global climate system. Warmer climates are now regarded as a serious global threat. The Intergovernmental Panel on Climate Change (IPCC) presented its third report on the climate in 2001 (IPCC (2001a), IPCC (2001b), IPCC (2001c)). The panel reports that the climate has warmed up by 0.6° C over the last century, while sea levels have risen by 10 to 20 cm on average. Further observations, such as the rise in precipitations along the intermediate and higher latitudes of the northern hemisphere, and the receding coverage with snow, glaciers and sea ice, complete the picture of global warming.

The answer to the question as to whether humanity does have an influence on the climatic phenomena observed is a clearer yes than in earlier IPCC reports. While the second IPCC report of 1996 offered the cautious statement that "the balance of evidence … *suggests a discernible human influence on global climate*", we read in the third report that *"most of the observed warming over the last 50 years is likely due to increases in greenhouse gas concentrations due to human activities."* For the period from 1990 to 2100, the IPCC forecasts a rise of the mean global temperature by 1.4 to 5.8° C due to increasing greenhouse gas emissions. Sea levels are predicted to rise by between 9 and 88 cm. Temperatures and sea levels will continue to rise beyond 2100, according to the IPCC, since greenhouse gases have a long atmospheric life.

Resource shortages and lifespans

Presently, the global energy system relies on fossil fuels at the rate of 83% (fig. A.2, app.1). From the perspective of critical sustainability, which allows, in principle, for the consumption of non-renewable resources within certain limits, questions arise about "safe minimum standards" for the consumption of fossil resources. The two main issues are (1) the lifespans left before non-renewable resources are exhausted and (2) their geographic availability, including price stability in connection with the latter. The analysis of these must take into account the inertia against changes of the energy system.

Concerning the first question, which is about the lifespan of certain non-renewable resources, we should note that past predictions regarding the limits of a specific resource, and the shortage preceding its exhaustion, always turned out to be problematic if not misleading. Still, it would be rash and dangerous to conclude that real shortages would only arise in the very distant future. The demand for energy resources has never been as intensive, and the time required for remodeling an energy system based on these resources has never been as long as it is today.

---

[74] 1 ppm = 1 part per million.

The uncertainty of any prognosis concerning the future development of resource supplies goes back to the fact that the volume of reserves of a specific resource is subject to wide fluctuations, for two reasons: Firstly, there is the possibility of new discoveries; secondly, due to price changes, known deposits can suddenly turn into reserves i.e. resources that can be extracted economically. Considering such interdependencies, every quantitative definition of reserves and resources of fossil energy carriers, as attempted in table 4.3, must be interpreted with due caution. Still, neither are these estimates completely arbitrary. This is shown in fig. C.6 in appendix 1, where the figures of tab. 4.3 (the rightmost bar in fig. C.6) are listed side by side with other calculations. All resources are expressed in terms of their energy content, so that the various fuels can be compared to each other and to the annual demand.

If this estimation of reserves is difficult enough, the calculation of the reserve lifespan of non-renewable resources suffers from the additional problem that the future development concerning the use of different resources is still unknown. Calculating the linear lifespan, i.e. the ration of reserves and present consumption as presented in table 4.3, at least enables us to make a relative comparison between different fossil energy resources.

According to table 4.3, the linear lifespans of mineral oil and gas are of similar length, whereas that of coal is four to six times longer. The linear lifespan of the known oil reserves is 40 years. Assuming an annual demand growth of 2% under otherwise constant conditions, the actual lifespan would fall to 30 years. In reality, shortages would lead much earlier to price rises and hence to resources turning into reserves and/or to a fall in consumption. The moment at which annual oil production can no longer be increased is thus even more important than the absolute

Table 4.3 Reserves and resources of fossil fuels compared to the present consumption. Figures in units of EJ (etajoule = 10^{18} joules) and EJ/year, respectively. In fig. C.6 (app.1), these figures (rightmost in the bar chart) are compared with those cited by other studies. Source: UNDP/OECD/WEC 2000.

	Reserves (EJ)	Additional resources (EJ)	Annual consumption 1998 (EJ/year)	Linear lifespan of reserves (years)[a]
Oil	6.000	6,100	147	40
added oil shale, tar sands, bitumen	*5,000*	*15,000*		
Coal	21,000	179,000	94	220
Gas	5,500	11,1000	84	65
added methane from coal mines and other sources	*9,400*	*24,000*		
Methane hydrate		*930,000*		

[a]   The linear lifespan is the ratio of reserves and annual consumption (assumed constant)

lifespan. The production from an oil field is represented by a characteristic bell curve. This also applies to the total of all oil fields and hence to the development of global oil extraction. The peak of the curve, i.e. the moment of maximum oil extraction is reached when about half of the existing oil has been extracted. It is unclear when this moment will arrive. Schindler and Zittel (2000) assert in their answers to questions posed by the Enquête Commission of the German *Bundestag* that, on the basis of the official, published data, this peak production will be reached between 2010 and 2015. There is ample evidence that the remaining reserves, especially in the countries of the Middle East, are overstated.

Another uncertainty concerns the size of undiscovered oil resources. The figures in table 4.3 are based on a doubling of the known reserves, although, according to some analyses, no significant oil discoveries should be expected in the future (fig. C.6, app.1). About 42,000 oil fields are known today. 75% of the mineral oil found to date lies in only 1 % of these oil fields. Most of the 400 largest fields were discovered more than 30 years ago (Schindler and Zittel 2000). In this context, fig. C.7 in appendix 1 is quite revealing: It shows that according to virtually all energy scenarios, there will be an acute oil supply gap between 2030 and 2050.

According to table 4.3, natural gas reserves would last for at least another 60 years. Nevertheless, demand is rising strongly, and the geographic availability is more limited than is the case for oil. The German Enquête Commission's comments in this respect (from Schindler and Zittel 2000, p. 22):

> Taking into account that mineral oil is substituted increasingly by natural gas, it appears extremely unlikely from today's perspective that cheap natural gas will still be available in sufficient volumes in 30 to 40 years. Considering the growth in the consumption figures, future discoveries are significant only insofar as they would delay a shortage of available gas by one or two decades at most, especially as the majority of future discoveries will be, most likely, in economically unfavorable deep offshore areas, far away from the consumer.

In principle, this situation concerning natural gas and methane, respectively, may change dramatically, if the methane hydrates, which are stored in deep-ocean sediments, could be extracted economically (see table 4.3). So far, however, the development of the technology needed to this end is at least as far from success as nuclear fusion reactor technology.

With coal, the situation is less dramatic: A the present rate of consumption, coal resources will last for more than another 200 years. Still, this lifespan would shrink rapidly if coal were to shoulder the contributions of the slowly depleting oil and gas resources. Furthermore, a switch to coal as the leading energy resource would significantly increase the CO_2 emissions per unit of energy and hence work against the desired decarbonization of the energy system. We will return to this point in chapter 4.3.1.

The geographic availability of fossil energy resources and the problem of political stability

The global lifespans of fossil energy resources (tab. 4.3) conceal considerable geographical variations in the consumption and, most notably, the location of the resources. This is the case for oil and natural gas, in particular, whereas coal is distributed more evenly over the continents (fig. 4.3). In fig. 4.4, the mineral oil resources are broken down further, geographically, and displayed in comparison to the total volumes extracted to date. The chart suggests that this unevenness in the distribution will continue to grow, in the medium term, since the countries showing the largest demand (e.g. USA) have already exhausted most of their domestic reserves. In the US, the extent to which the demand is met by domestic resources has fallen continuously over the last 30 years (fig. 4.5), whereas not many years before the first oil crisis in 1973, domestic resources still contributed nearly 80%. How much more vulnerable the American economy must be today, as that figure has dropped to 40% of a higher total demand.

This observation illustrates impressively the present risks arising from the regional concentration of oil and – to a somewhat lesser extent – gas in the Arab Gulf States. The intensified conflict surrounding Palestine, and the international tensions in connection with the war against terrorism, brought into focus the specific risks. The possibility of renewed military conflicts, but also the internal instability of autocratic regimes, renders the assumption of a long-term security of supply from that region rather improbable. Even more than the first gulf war of 1991, the overthrow of the Shah of Iran in 1979 showed how such events could shock the world economy. Since we will become more dependent on oil from the Middle East again, the "multiplicator effect" [of any problems arising in that region] – and hence the temptation to hit the "soft underbelly" of the western industrial nations – will also increase.

Social aspects and the north-south conflict

Apart from the uneven geographical distribution of fossil energy resources, the unbalanced utilization of these resources is an issue in the context of sustainability (see fig. 4.1). As already discussed, countries with a per-capita consumption of markedly less than 1,000 watts are hampered in their development (Goldemberg et al. 1985). Large sections of the population have virtually no access to commercial energy in those countries. This applies to electricity, in particular, which becomes ever more important for economic and social development in the age of information and services.

Excursus: Nuclear fission and fusion as backstop technologies?

Following the first oil crisis of 1973/74, nuclear energy in particular – with fully developed breeder and reprocessing systems – was assigned the role to overcome the limitations and (the then political) risks of fossil energies and make available, in an autarkic fashion and at low prices, whatever amount of energy is needed. The use of this backstop technology was also thought to check the negotiating power of the OPEC cartel. Hence, the third energy programme of the German federal government made provisions for constructing four large nuclear power plants per year, in addition to massive investments in coal gasification and liquefaction. The renowned "Häfele Studies" predicted the existence of 200 breeder reactors in 2020 – in India alone. The real development, however, crushed this utopia in various ways:

– In the US and northern Europe, massive acceptance problems led to long delays in the licensing process, costly retrofits and political risks concerning decom-

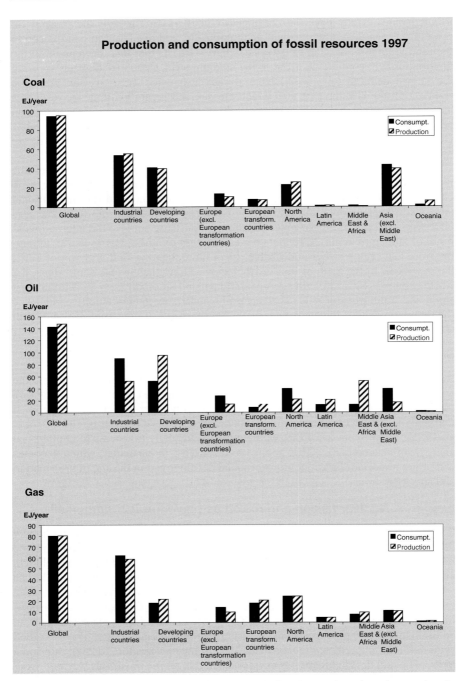

Fig. 4.3 Consumption and production of coal, mineral oil and natural gas in various regions in 1997. Data source: WRI (2001).

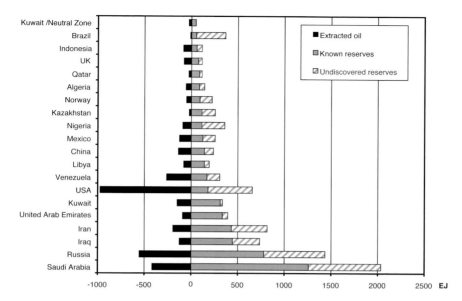

Fig. 4.4 Production to date and remaining reserves of the most important oil producing countries. Many countries, including the USA, have already extracted most of their reserves of mineral oil. Only few still have the potential to extend their oil production, and nearly all of these are in the Middle East. Data source: IEA (2001a).

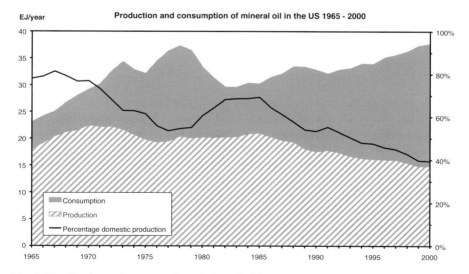

Fig. 4.5 Production and consumption of mineral oil in the US 1965–2000, in EJ per year, and the proportion of domestic production in percent. Data source: BP (2001).

missioning – while the growth of the electricity market was flatter than expected back in the mid 1970s. Furthermore, the issue of the ultimate disposal of radioactive waste turned out to be more difficult than expected, both politically and technically.

– Control of proliferation risks through the International Atomic Energy Agency was difficult, too. Iraq, North Korea and, probably, India, Pakistan, South Africa and Israel used elements of the "civil" nuclear fuel cycle for military purposes. It is unclear to what extent terrorist organizations , perhaps supported by so-called 'states of concern' (formerly 'rogue states'), have also obtained nuclear technologies.

– Political protests and deregulation thwarted "centralistic" nuclear energy. Facing competitive markets and increased capital costs, preproduction investments of ca. € 4 bil. and construction and commissioning periods of ten years could not be justified anymore, economically. In the UK, for instance, the deregulation that took place in the 1980s halted any further investment in nuclear energy, regardless of the very "pro-nuclear" stance of the then Conservative government. Existing nuclear power plants could only be privatized with massive write-offs (if at all).

The climate risk, and the political risks, connected with oil have revitalized the discussion about using the CO_2-free and quasi-domestic energy from nuclear fission. Concepts for small, inherently safe high-temperature reactors, which do not present any proliferation risks, are attracting (renewed) attention. However, we do not have to decide here whether there will be a renaissance of nuclear energy with completely new technological concepts. For any realistic estimate concerning a timetable for the resumption of nuclear fission technology suggests that a significant contribution to global energy production could not be expected before 2050. The political decisions concerning energy, however, must be made much earlier. In fact, the present situation, at least in Europe, appears to be that the importance of nuclear energy is rather in decline than rising. By about 2030, all European nuclear power plants must be replaced, because their licenses will have expired, which leaves us with another challenging question: how to implement their replacement.

We see a similar situation in the field of fusion energy. Even if there has been significant progress – especially with regard to the service life of magnets and the confinement time of the plasma –, we are still decades away from any commercial utilization. Therefore, the fusion reactor and the further development of new fission reactors remain options for basic research. They will contribute neither to lowering CO_2 emissions nor to enhancing the security of procurement before 2050.

4.2.5
Operationalizing critical sustainability: the "time of safe practice"

Cultural history teaches us that all practices of human society are pursued for a limited time only. Hence, the question as to whether non-renewable resources are used cannot *on its own* constitute a meaningful criterion of sustainability. This justifies the choice of critical sustainability as the objective to aim for. The *concept of the time of safe practice* opens up the possibility of implementing this objective. It assumes that every social activity can be analyzed with regard to how long it can be continued unchanged before it approaches its own limitations. It is this time span that we refer to as *the time of safe practice* (Imboden 1993). Concerning the use of non-renewable resources, it is equivalent to the *lifespan* cited in tab. 4.3. Similar considerations can be applied to other aspects. The extent of cultivated surface areas disappearing due to construction activities, for instance, provides the basis for calculating the time after which a significant portion of agricultural land will have been used up.

In aggregating the individual times into a "total" time of safe practice and thus assessing the system as a whole, the shortest times play a dominant role. This may be illustrated by a comparison with the analysis of a road network: Knowing the annual growth of traffic in various locations and the respective road capacities, one can calculate for how long the individual sections will be able to cope with the traffic. Traffic planning will particularly focus on those stretches of road for which the (critical) times thus calculated are shortest. Even if, due to possible shifts in traffic distribution, these times are not independent (neither are the various times of safe practice independent, because of substitutions and other measures), one can still analyze, in first approximation, the vulnerability of the system as a whole on the basis of the shortest times.

There are, of course, not only mechanisms that shorten the time of safe practice, but also those that lengthen it. The exploration of new resources, the substitution of one resource with another that has a longer lifespan, or the creation of new jobs are examples of the latter. Where a society managed to survive for a long time – for instance the culture of ancient Egypt – , it must have followed a strategy that always kept the time of safe practice from falling below a critical threshold or, in other words, ensured that the rule of critical sustainability was never violated.

Now we can quantify critical sustainability by means of the concept of the "time of safe practice", from this perspective. To this end, we consider, firstly, the *change* of sustainability and assert the following, plausible postulate:

1. *A practice (e. g. an energy practice) is sustainable if the time of safe practice is constant or growing (principle of the constant time of safe practice).*

Applied to the use of non-renewable energies, this means that the substitution rate, and the rate of discovery of the new resource, should at least counterbalance its consumption. Without new discoveries, the consumption of a resource of, for instance, 40 years' linear lifespan (corresponding to the linear lifespan of mineral oil in table 4.3) would have to be reduced by the factor 1/40 i.e. 2.5%, annually, in order to keep constant the time of safe practice. Concerning the issues surrounding CO_2, the principle of a constant time of safe practice means: The decarbonization rate of the global energy system is sufficiently high that – notwithstanding the growing energy demand – CO_2 emissions into the atmosphere do not exceed a figure given by the climate models and determined according to certain normative criteria. We will return to this point at a later stage (see tab. 4.6 and fig. 4.7).

A second requirement of sustainability is based on the changeability or inertness of a system, e.g. a national energy system. The inertness can be defined as the time needed for a significant change in the system concerned. For the present energy system, such a significant change could result in, for instance, the substitution of the present fossil energy supply by renewable energy resources or by nuclear energy. Therefore, the second postulate of sustainability reads:

2. The time of safe practice must exceed the inertness of the system concerned.

Table 4.4 CONSTRUCTED SWITZERLAND (CS)

The "Constructed Switzerland" comprises ca. 2 mil. buildings, which are connected by a complex infrastructure (roads, railways, water supplies and wastewater disposal, energy distribution, telecommunication etc.).
The replacement value (in bil. CHF) of the CS in 1999 amounted to: [a]

Buildings	1,788
Construction excl. buildings	653
Total replacement value	**2,441**

Annual expenses, in bil. CHF, for new constructions, conversions and renovations in Switzerland in 1999[b]

New constructions	23
Conversions	14
Public maintenance works	3
Total expenses per year	**40**
Theoretical renewal rate	1.6 % per year, (corresponding to a renewal time of about 60 years)

The BWS involves:

direct: ca. 60% of the total consumption of final energy

indirect: a large part of the energy used for mobility, which in turn amounts to ca. 30% of the total consumption of final energy

---

[a]   Wuest & Partner 1999.
[b]   BFS 2001.

The inertness of a system determines the transition potentials from a non-sustainable system to a sustainable one. Inertness itself is determined by innovations and the economic potential, on the one hand, and by certain structural features of a country or region, on the other. Considered in a historical context, socio-economic and political structures and traditions constitute circumstances that lead to inertias. Apart from these factors, and viewed from a more technical/scientific perspective, the *Constructed World* (or the national Constructed Germany, Switzerland etc.) is the most important factor contributing to inertness against changes in the energy system. This inertness is shown in tab. 4.4, using *Constructed Schweiz* as an example.[75] As indicated above, *Constructed Switzerland* is understood as the sum of the ca. 2 million buildings and the entire infrastructural network (roads, railways, water supplies and disposal, energy distribution, telecommunications etc.). The total replacement

---

[75]   This example is not supposed to suggest that the edifice is he only factor determining the sustainability of a country. Nevertheless, it is an important factor of influence, particularly since about 60% of the final energy consumption in Switzerland is directly connected to the construction, running and maintenance of the edifice.

value of CS amounts to about 2.4 trillion Swiss francs. The annual expenses for new buildings, conversions and renovations total 40 billion francs. Hence, the theoretical renewal rate of the CS is 1.6 % per year, equivalent to a mean renewal time of about 60 years. Consequently, its conversion would take two to three generations. The renewal time of the CS in relation to energy demand could be reduced by one to two decades by giving preference to energy-relevant measures in its conversion. (The conversion investments amount to 35% of the total expenses for the CS.)

The time of safe practice and the inertness of the system constitute the quantitative basis on which the energy system can be analyzed with regard to critical sustainability. The comparison between the lifespans of fossil energy resources, on which today's commercial energy system is mainly based, and the renewal times of the edifice of a typical industrial country such as Switzerland leads to the conclusion that the sustainability of this system is inadequate and fragile. The inertness of the energy system makes the economy sensitive to volatile price fluctuations such as those occurring increasingly during the last 30 years. Energy forecasts indicate that the global energy system is not developing towards increased sustainability, if the present trend continues. There is, rather, the danger that the problems pointed out will become even more severe. A re-orientation of energy policies is unavoidable. Hence, in a world full of political obstacles, a reference point or benchmark is called for. In the next section, we will try to outline the essential cornerstones of a future, sustainable energy system.

4.3
Reference points for the sustainable supply of energy on a global scale

4.3.1
Options for change

Apart from changing one's personal objectives, which is possible through changing predominant lifestyles (practical discussion of sustainability), there are two options with regard to the choice of instruments: the *substitution* of energy carriers, and improving the *efficiency* of energy use (theoretical discussion of sustainability, see section 3.3 above).

Innovations play a central role in the practical as well as in the theoretical discussion. In the theoretical discussion, we deal with rather technical issues, whereas the practical discussion is more about common questions (e.g. reading a book instead of going for a ride in the car). On the one hand, technology is easier to plan and implement, because the consumer, as a factor, can be largely bypassed by using intelligent technologies. On the other hand, the technical solutions are limited by certain unshakeable boundary conditions, for instance the restrictions imposed by physics or the geochemical and biological make up of the earth. Exactly the opposite is true for the practical discussion: Lifestyle changes cannot be planned or decreed through technical or organizational measures from above; any such attempts by dictatorial regimes and planned economies eventually failed in every case. Activities and events, on the other hand, can change the world very quickly. The 9-11 terror attacks on the US in 2001, for example, dramatically changed the development of air traffic, at least for the short term. No government measure would be likely to achieve such a radical transformation.

Substitution aims at reducing CO_2 emissions into the atmosphere as well as effecting a shift from non-renewable to renewable energy carriers, and hence, in general, from "insecure" oil and gas sources to a more local energy production. Chapter 5 provides an overview of the global energy resources, which are available, in principle, for substituting fossil fuels; we also discuss which potentials need to be realized. Without going into the details of that discussion, we can draw some simple conclusions for the potentials and boundary conditions of the future energy system:

1. From a purely technical perspective, the potential of the (non-renewable and renewable) energy resources is sufficient to meet the global energy demand for the coming centuries.

2. The use of non-renewable energy resources will encounter limits – resulting from the climate issue, in the case of the fossil fuels, and because of economic and political problems surrounding nuclear energy –, which will become noticeable within the next 50 years. Still, due to the inertness of the global energy system, this poses the risk that, in spite of the Kyoto Protocol, societies will continue to pursue the fossil path.

3. The total potential of all solar-based renewable energies (including hydroelectric power and biomass) may suffice to meet the global demand, but as long as the latter rises by 2% annually, there is little chance for solar energy to increase its present, marginal share of less than 1% to the necessary extent over the next 50 years. This would require corresponding efficiency improvements, which, especially during the earliest decades, are more important in volume than the contributions from regenerative energy sources.

4. The potential of technical efficiency improvements is far from exhausted. In the construction sector, in particular, which accounts for more than 50% of the energy demand in many countries, efficiency can still be increased by a factor between 3 and 10. In the mobility sector, too, improvements by a factor of at least 2 to 3 are possible. Overall, the unexploited potential is found in private consumption, predominantly, and to a lesser extent in the manufacturing industry, where economic considerations have already led to appropriate measures.

The quintessence of the above considerations can be summarized in the following simple calculation:

An energy path that aims at meeting the conditions of critical sustainability (i.e. one that involves a massive reduction of CO_2 emissions) while relying on the reduction of carbon intensity alone faces an enormous technical challenge: For a GDP growth of 2% per annum, as targeted by the western industrial countries, and a concurrent annual reduction by 2% demanded for CO_2 emissions, CO_2 intensity would have to fall by 4% per year.

We will specify, in the following, the benchmark to be achieved by such a development by 2050, before we examine further the technical innovation potentials that may enable the achievement of that benchmark.

4.3.2
The 2000-Watt benchmark: sustainable comfort through intelligence

A sustainable energy system of the future must rest on two pillars: (1) the intelligent use of energy and (2) the increased use of solar resources. The central question is what amount of energy is needed by an intelligent person in the future without having to forsake the western standard of living enjoyed today. Goldemberg et al. (1985) estimated a demand of 1,000 Watt per person in the developing countries. For the affluent countries of the west, where there is a higher demand for heating and other practices in connection with individual mobility, this target would certainly be too ambitious. On the other hand, the typical demand shown by the wealthy countries today (4,000 to 10,000 Watt per person) is, most probably, too high. The solution must be somewhere – where exactly cannot be determined unequivocally by scientific methods alone – in the middle. Therefore, a sustainable energy target is also a normative one, at least to a certain extent.

We point to a project at the *Swiss Federal Institute of Technology* in Zürich (ETH), which was developed in 1998, under the title "The 2000-Watt Society". It is based on the conclusion explained above, according to which a country like Switzerland would be able to manage with 2,000 Watt per person, without any loss in the living standard (Imboden/Roggo 2000). In fig. 4.6, today's energy demand, divided into different sectors, is compared to the demand at the 2000-Watt benchmark. Virtually the entire savings potential of some 60% is in the heating ["home and work"] and mobility sectors.

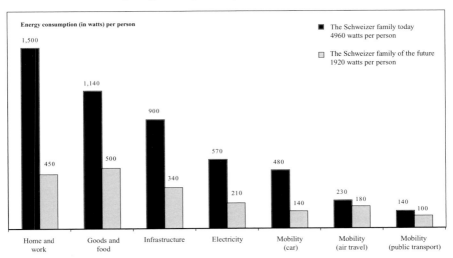

Fig. 4.6 Taking a fictitious, but typical family as an example, the bar chart shows the activities for which the Swiss family uses energy today and in the "2000-Watt society" of the future. The potential for reduction lies mainly in the areas of "home and work" and "mobility". Only technologies that are available today are taken into account. (Imboden and Roggo 2000).

The 2000-watt benchmark can be justified not only from the consumption side but also from the supply side. One third of primary energy demand could be met by fossil fuels and two thirds by solar energy, without adverse effects for humanity and nature in a future world accommodating about 10 billion people. Table 4.5 shows that using only 700 to 1,100 Watt per person from fossil energy resources in 50 years and just 250 to 450 Watt person from such resources in 100 years would be entirely compatible with the objectives of climate protection. On the other hand, developing a potential for renewable energies of 1,500 to 1,700 Watt per person over the coming 100 years would be a realistic project. In Switzerland, 600 Watt per person is available from hydroelectric power already. The remaining demand can be met by other renewable sources (photovoltaic energy, wind, thermal solar energy, and biomass) in the future. However, it would not be realistic to assume the possibility, in the foreseeable future, of providing 10,000 Watt of power per person based on renewable energy resources alone.

The figures shown in tab. 4.5 are based on the so-called IPCC S450 scenario, according to which the atmospheric CO_2 concentration should not exceed 450 ppm. This is not an arbitrary figure. It arises, firstly, from the natural variation range of the mean global temperature over the last 100,000 years, which molded our present environment and can hence be used as a benchmark for the maximum ensured adaptability of ecological and social systems. The maximum mean temperature in the younger Quaternary (Eem warm period) was 16.1°C; the upper limit for an acceptable global temperature can be assumed to be at 16.6°C, with a variation margin of + 0,5°C. Today, we are only 1.3°C below this "crash barrier" (WBGU 1997, S 15 und WBGU 1995, S 8).

Secondly, the adaptability of ecosystems depends, crucially, on the rate of change. Again based on the natural climate history, 0.1°C per decade was defined as the maximum warming rate that would be just manageable in ecological terms (see WBGU 1997, p. 16, and Onigkeit et al. 2000, p. 60). In fact, to avoid risks, we should strive to stabilize CO_2 concentrations at the present level of 365 ppm (S350-Szenario). However, this would require an immediate reduction of 50% of the relevant emissions. Such a radical change would be impossible to implement, both economically and politically. In ecological terms, the slightly less stringent S450 scenario appears more promising, in this respect, and should about meet (as far as this can be assessed at all, considering the substantial uncertainties involved) the historic "demands" on the maximum allowable temperature rise and rate of this rise, respectively.[76] If the atmospheric CO_2 concentrations were to be kept at the level of 450 ppm, the models predict a probable global temperature rise of ca. 1°C by the year 2100, compared to the figure for 1990.[77] This would imply, at the same time, a sea level rise by ca. 30 cm by the end of this century (Onigkeit/Alcamo 2000, p. 79); the most recent computations even suggest a sea level rise by about 80 cm. This prognosis is only valid under the proviso that the emissions of methane, dinitro oxide and sulfur oxide stabilize at the level of 1990 and the ozone figure follows the path described in the Montreal Protocol.

---

[76] In this scenario, the temperature rise per decade, up to the year 2030, exceeds 0.15 °C; it would not fall below 0.1°C before the second half of the century.

[77] Compared to the preindustrial figure, the temperature rise computed by Onigkeit/Alcamo (2000) constitutes an increase of 1.7 °C.

Table 4.5 The global CO_2 situation in the IPCC S450 scenario

Present CO_2 emissions

Total	23	gigatons (Gt) CO_2 per year
Global average	4	t CO_2 per year and person
USA	20	t CO_2 per year and person
OECD countries	11	t CO_2 per year and person
India	0.9	t CO_2 per year and person

Total CO_2 emissions for the S450 scenario

Sum over the period 1990 to 2100		2,500 Gt CO_2
for comparison (see tab. 4.3)[78]		
burning all	oil reserves	440 Gt CO_2
	coal reserves	2,000 Gt CO_2
	gas reserves	310 Gt CO_2

Allowable CO_2 emissions for the S450 scenario

	Population (bil.)	Total emissions (Gt CO_2/year)	Emissions per person (tCO_2/year)	Coal	Oil	Gas
				Potential for energy use[79] (watts per person)		
2050	10	20 (15 to 40)	2.0 (1.5 to 4.0)	700	900	1,100
2100	12	10 (7 to 18)	0.8 (0.6 to 1.5)	250	350	450

The models offer various options concerning the time frame of the reduction (see the emission corridors in WBGU 1997). The later the measures set in, the more cost-intensive they will be, and the more limited the options for action of future generations. For the S450 scenario, the cumulative CO_2 emissions from 1990 to 2100 must not exceed 2,500 Gt; after 2100, the emissions should stabilize at a level of about 10 Gt CO_2/year. The present figure is ca. 24 Gt CO_2/year. Consequently, a global energy system meeting the 2000-Watt benchmark would cover 20% to 30% of the demand by using fossil energy sources; the remainder would come from solar energy. With the human population estimated to be 10 billion at that point, the global energy demand would amount to 20 TW, or 630 EJ per year, which is nearly double the present figure. The large discrepancies between wealthy and poor countries would have largely disappeared.

Most energy forecasts paint a completely different picture (see appendix 1, section C). In most forecasts, fossil fuels will continue to be the most important pillars of the energy system. The absolute growth in consumption would still be dominated by the affluent countries, meaning that the discrepancies between rich and poor

[78] Computed with the following CO_2 emission coefficients (in kg CO_2/GJ): coal 94.6, crude oil 73.3, natural gas 56.1.
[79] Figures valid for the most likely emission quotas per person; emission coefficients as above.

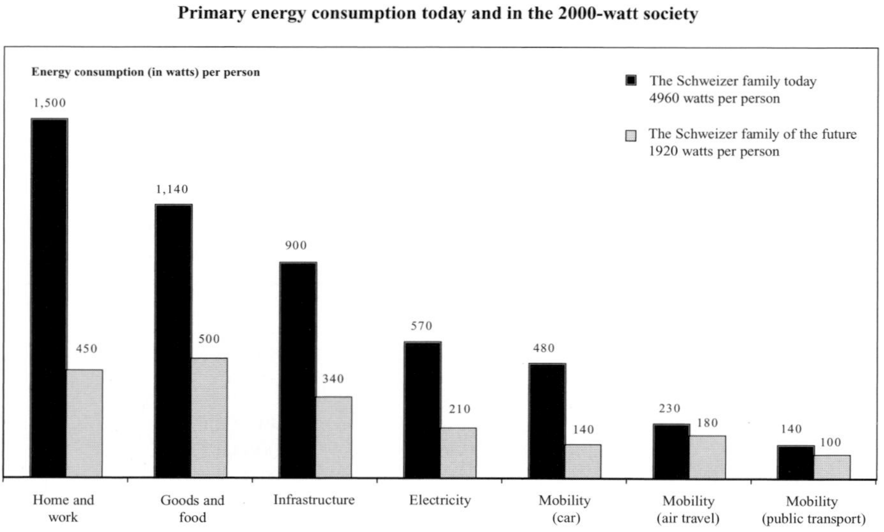

Primary energy consumption today and in the 2000-watt society

Energy consumption (in watts) per person

■ The Schweizer family today
4960 watts per person

☐ The Schweizer family of the future
1920 watts per person

1,500
1,140
900
570
480
450   500
340   210   230   180   140
140   140   100

| Home and work | Goods and food | Infrastructure | Electricity | Mobility (car) | Mobility (air travel) | Mobility (public transport) |

Fig. 4.7 Cumulative CO_2 emissions between 1990 and 2100 for various energy scenarios (see appendix 1, section C). With only one exception, 2,500 Gt CO_2 – the figure allowed in the S450 scenario – would be exceeded by 2100.

countries would be even more pronounced. The atmospheric CO_2 concentrations would not be stabilized. With few exceptions, 2,500 Gt CO_2 – the maximum cumulative input into the atmosphere given in the S450 scenario – would be exceeded before 2100, without any stabilization in sight (fig. 4.7). The divergence between trend and benchmark demonstrates the urgent need for political, economic and social action, and the enormous challenge to the human capacity for innovation.

In chapter 5, we will look at the technical potentials for closing this gap by improving energy efficiency and using regenerative energies.

5 Potentials for the Sustainable Development of Energy Systems

5.1
Introduction

As we have seen in the previous chapter, the use of fossil energy is a source of environmental damages and risks. Energy use in the European Union is 4.5 kW per capita (see Fig. B1, Appendix to chapter 4), which is beyond levels that can be called sustainable. Of the environmental risks, the possibility of climate change through the emission of carbon dioxide is a prominent one. Greenhouse gas emissions in the European Union are 10.5 t carbon dioxide equivalent per capita per year. According to the present understanding, emissions of carbon dioxide need to be strongly reduced, at least in industrialized countries.

There is already a wide range of technologies available to reduce energy-related emissions of carbon dioxide. Existing technologies can improve the efficiency of energy production and consumption by 20–40% in one or two decades. Important options are, for instance, the extension of the application of building insulation, the use of building management systems, the optimization of electric appliances, the application of combined generation of heat and power (CHP), the further improvement of car engines and aerodynamics, and a large set of adaptation options for industrial processes. Another option is a shift from coal to natural gas. Also, renewable energy sources can already play a limited role in the coming decade. Such options are available and are sufficient to reach short and medium term targets, like those set by the Kyoto protocol for the period 2008 to 2012 (K. Blok, D. de Jager, C.A. Hendriks: Economic Evaluation of Sectoral Objectives for Climate Change – Summary for Policy Makers, European Commission, DG Environment, 2001).

Reaching longer term targets, like those for a period of 30–100 years ahead requires not only the further adoption of the technologies mentioned, but also the application of new technologies. These technologies should be adopted at a rate that at least compensates for the increase in human activities associated with economic growth. The aim of this chapter is to investigate whether a more sustainable energy system is feasible. To this end, technologies that are important for the longer term on the road towards a more sustainable energy system will be evaluated and their possible role for the European Union will be quantified.

Production	Conversion	Transport Storage Distribution	End-use conversion	End-use
e.g. coal mining wind turbines natural gas prod.	e.g. refineries power plants	e.g. oil storage electricity grids	e.g. boilers combined-heat-and-power-plants	e.g. space heating cooking lighting

Table 5.1 Schematic description of the stages of an energy system (see also chapter 2, Box 3)

An energy system can schematically be broken down into five stages (see table 5.1). Energy is produced from natural resources, often converted into other energy carriers and subsequently transported and distributed to the so-called final consumers. The amount of energy that is used by final consumers is called final energy use. The final consumers can convert the energy further and finally use it for a wide range of applications.

In this paper, we will focus on two categories of options: energy efficiency improvement and substitution (see chapter 4). First, we will discuss energy and material efficiency improvement options that exist in end-use and conversion (section 5.2). Next, renewable energy options will be covered (section 5.3). Note that the aim of this paper is just to present some overall lines, illustrated with some specific examples. More extended analysis can be found in the literature (K. Blok, W.C. Turkenburg, W. Eichhammer, U. Farinelli, T.B. Johansson (eds.): Overview of Energy RD&D Options for a Sustainable Future, Office for Official Publications of the European Communities, Luxembourg, 1996. Federal Energy Research and Development for the Challenges of the 21st Century, President's Committee of Advisors on Science and Technology (PCAST), Washington D.C., 1997).

On the basis of these analyses, some images are presented and the prospects for a sustainable energy system in the long term will be discussed (section 5.4). Finally, an overview will be given of barriers for the materialization of the technological potentials (5.5).

5.2
Technical energy efficiency improvement

Traditionally, the energy efficiency of energy use improves by about 0.5–2% per year, with the higher rates occurring in periods with high fuel prices or active energy policies. In this section the feasibility of enhanced energy efficiency improvement will be discussed.

First of all, some more need to be said about what is meant with technical energy efficiency[1] improvement. Technical energy efficiency is output divided by input of

---

[1] Here the attribute 'technical' is added to distinguish it from the (overall) energy efficiency defined in Chapter 4 (Box 4.1) as GDP per energy consumption of a country. Where the context is clear, the adjective 'technical' is omitted.

a process in which energy is an input. The opposite of energy efficiency is input divided by output, often indicated as the specific energy consumption. If this chapter talks about energy efficiency improvement, always a decline of the specific energy consumption is meant.

In this section, opportunities for technical energy efficiency improvement in the three major end-use sectors: manufacturing industry, residential and commercial buildings, and transportation will be discussed.

Manufacturing industry

The opportunities for future energy efficiency improvement for a number of sectors in heavy industry were analyzed by De Beer[2]. This was done according to a structured method, including: (i) process and energy analysis; (ii) technology identification; (iii) technology characterization. The results of this work provide for an overview of the potential for energy efficiency improvement for selected sectors in manufacturing industry, see Table 5.2. We see that in all the cases it is possible to bridge about half of the gap between the present best technologies and the thermodynamic minimum with identified new technologies. Despite the extended inventories that were made, all identified technologies could be commercialized well within 30 years, and the overwhelming majority even within 15 years. However, actual development of new industrial process technologies is a slow process (E. Luiten: Beyond Energy Efficiency, Ph.D. Thesis, Utrecht University, 13 September 2001); autonomous development alone will not lead to the development of these technologies within these timeframes.

It is clear that in a number of these cases the further decrease in specific energy use is limited by the thermodynamic minima that exist for certain conversions. However, in addition the energy demand for producing these materials can be further limited by using primary materials in a more efficient way. This can be done through:

− more material-efficient product design;
− material and product recycling;
− material cascading;
− material substitution, including the use of biomass-derived materials.

Some studies have shown that with existing technology substantial improvement in material efficiency is possible, both for individual products ('eco-design') and for integral material systems. However, long-term prospects are still unknown and certainly not quantified. It is not unlikely that – as is the case with energy efficiency – also in the case of material efficiency further innovations seem possible, including the development of new materials; alternative inputs and processing routes for existing materials; tools for the development of material-efficient products; and improved material recycling through the use of material recognition systems, better separation techniques and new logistic systems.

---

[2]   J. de Beer: Potential for Industrial Energy-Efficiency Improvement in the Long Term, Ph.D. Thesis, Utrecht University, 1998. Parts of this thesis are also published as: J. de Beer, E. Worrell, K. Blok: Future Technologies for Energy-Efficient Iron and Steel Making, Annual Review of Energy and Environment, **23**(1998)123-205; and: J.G. de Beer, E. Worrell en K. Blok, Long-term energy-efficiency improvement in the paper and board industry, Energy, the International Journal, 23(1998)21-42.

Table 5.2 Overview of present best technologies, and identified potential for improvement in terms of specific energy consumption (in GJ/t) for some industrial energy functions.

	Specific energy consumption levels (GJ/t)			Relevant future technologies
	Present best technology	Thermo-dynamic minimum	Combination of best identified future technologies	
Paper/board (paper drying)	2.3–8.6	0.0	0.6–4.3	Impulse drying Condebelt Dry sheet forming Airless drying
Primary steel production	19.0	6.6	12.5	Smelt reduction Strip casting
Secondary steel production	7.0	0.0	3.5	Combination shaft furnace Strip casting
Ammonia production	33.0	24.1	28.6	Membrane reactors
Nitric acid production	26.8	3.2	15.3	Gas turbine or solid-oxide-fuel-cell integration

Residential and commercial sector

Space heating and hot water production are important energy functions in buildings and responsible for about two thirds of primary energy demand. The developments for these energy functions can be illustrated by the developments for the residential sector in the Netherlands where the average energy use for space heating and hot water production was about 100 GJ (~ 1 GJ/m^2) per year in the late seventies, both for the average stock and for new buildings. At present the energy use for the average stock has declined to 70 GJ per year (R.J. Weegink: Basisonderzoek Aardgas Kleinverbruikers BAK 1997, EnergieNed, Arnhem, 1998). For new dwellings a standard was set in 1996 that forced building developers to build houses with a projected consumption of about 44 GJ. This was further decreased in two steps to 32 GJ in the year 2000.

Meanwhile, five real estate developers built a total of 200 buildings reaching a level of 19 GJ (completed in 1999–2000). This was achieved by expanding insulation, the application of heat recovery systems and the use of solar hot water heaters.

This is not the end of the possibilities. New technologies can be developed, like:

– the use of new insulation materials (notably the use of vacuum insulation);
– further reduction of the heat loss through windows;

– the application of heat pumps (instead fuel cells may play a role as heat source, also[3]);
– compact energy storage systems (making solar space heating possible).

In the end this technical developments could make it possible to build houses with zero energy use at affordable costs.

Transportation

Passenger cars are responsible for about two-thirds of energy use for transportation. During the nineties the energy use of passenger cars in the European Union was about 7 – 8 liters per 100 km (both for new cars and for the average). The prospects are good for a further decrease of specific energy use (see Table 5.3). Some Japanese manufacturers are already on the market with hybrid vehicles, combining a conventional engine and an electric motor. A step further would be the use of proton-exchange-membrane (PEM) fuel cells in cars, although it might be that the optimized hybrids will prove to be more efficient at the end (M.A. Weiss, J.B. Heywood, E.M. Drake, A. Schafer, F.F. AuYeung: On the Road in 2020, Energy Laboratory, MIT, Cambridge, MA, USA, 2000). There are even suggestions that applying new light materials, in combination with the new propulsion technologies, may bring fuel consumption levels to about 1 liter of gasoline-equivalent per 100 km (E. von Weizsäcker, A.B. Lovins, L.H. Lovins: Factor Four, Earthscan, London, 1998).

Table 5.3 Specific energy use of passenger cars (liter gasoline-equivalent per 100 km)

Present average in Europe	7–8
European standard 2008 (average new)	5.8
Hybrids on the market	4–5
Improved hybrids or fuel cell cars	2–3
Ultralights	0.8–1.6

Energy efficiency improvement of energy conversion

Energy use and energy conversion exists in many parts of the energy supply system. The most important conversion losses occur in the electricity sector, where world-wide only about one third of the fossil fuel input is converted to electricity, with a typical conversion efficiency of 40% in the most efficient countries.

Much higher efficiencies are possible already. Natural-gas fired combined-cycle power plants can be built with a conversion efficiency of nearly 60% by now, whereas the best coal-fired power plants reach somewhat more than 45%. The energy efficiency increase for natural-gas fired systems was most prominent in the past decades, mainly caused by a tremendous performance increase of gas turbines.

---

[3]   These options are not better than the present best technology (combined-cycle district heating) but more widely applicable, see: M.E. Ossebaard, A.J.M. van Wijk, M.T. van Wees: Heat Supply in the Netherlands: A Systems Analysis of Costs, Exergy Efficiency, CO_2 and NO_x Emissions, Energy 22(1997)1087-1098.

The expectation is that further improvement is possible, but the limitations are in sight: the maximum theoretical efficiency of an energy conversion process based on combustion is 70–75%.

The best-known example of an electricity production process not based on combustion is the fuel cell. Fuel cells are under development since many decades, but for stationary applications they have been 'locked out' by the rapid development of the gas turbine technology. It might be that, for instance, solid oxide fuel cells in combination with combined cycle plants may be able to reach conversion efficiencies above 70%[4].

The effect on carbon dioxide emissions can be even more pronounced than those on efficiencies. In 1995, the average emission factor for fossil-fuel based power generation in the European Union was 790 g CO_2/kWh (Derived from IEA Energy Statistics). For natural-gas based power generation with an efficiency of 70% the emission factor drops to 290 g CO_2/kWh.

Overall effects of enhanced development of energy efficient technology

The sectors of energy consumption discussed here together cover about half of the energy use in most industrialized countries. For each of these sectors we have identified options that make it possible to decrease specific energy consumption levels for *new* equipment at substantial rates, i.e., 5% per year or more. Although a 5% per year decline for new equipment is very substantial (if maintained for 50 years, this would be a decline by more than 90%), the effect on the *average* energy efficiency is limited due to the slow turnover of capital stock.

To analyze this effect a simple vintage simulation model was developed. In all cases the growth of the stock of energy using equipment is assumed to be 2% per year. In the reference case, an energy efficiency improvement of 1.5% per year for *all* equipment is assumed, leading to an increasing energy demand (see fig. 5.1, dotted curve). The reference case is compared to simulations in which an enhanced rate of energy efficiency improvement of 5% per year for *new* equipment is assumed for 80% of the energy applications (which, in fact, corresponds to a mean rate of 4.3% per year for all the applications). No effect of retrofit of equipment is taken into account, as this is relatively unimportant for the long term.

First, the effect of the average service life of the replaced equipment is examined:

– 15 years: typical for cars and household appliances;
– 30 years: typical for large-scale industrial process equipment and power plants;
– 60 years: typical for buildings.

The results are depicted in figure 5.1. The baseline shows an increase in energy use by about 25% in 50 years. In the enhanced energy efficiency improvement cases the energy use *decreases* by approximately 50% until 2050, except for the equipment with a service life of 60 years, for which the energy use decreases by only one third.

---

[4]   PEM fuel cells may be important in transportation or small-scale combined generation of heat and power, see section 2.

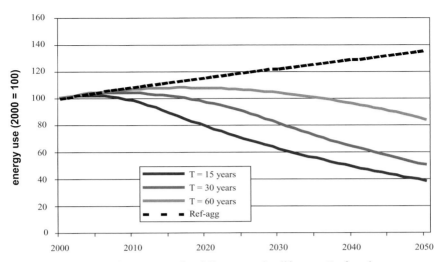

Fig. 5.1 Development of energy use for different service lifespans T of equipment, assuming that since 2000 the specific energy use decreases by 5% per year for 80 % of the new equipment. In the reference case the improvement rate for all equipment is 1.5% per year (dotted line). In all calculations an increase of energy using equipment of 2% per year is assumed.

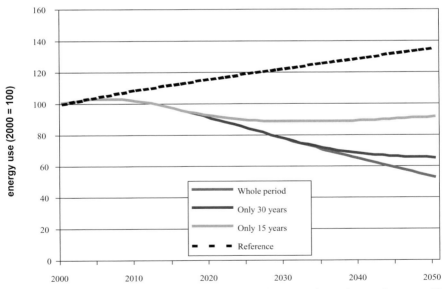

Fig. 5.2 As in fig. 5.2 with the additional assumption that the enhanced rate of energy efficiency improvement will only materialize for a limited period (i.e. for 15 years, for 30 years and for 50 years). The equipment has a mix of lifetimes (50% 15 years, 25% 30 years, 25% 60 years). Reference case (dotted line) as in fig. 5.2.

Next, it is assumed that the rapid rate of innovation can be maintained only for a limited period, i.e. 15 or 30 years. The results given in figure 5.2 show that even then the long-term effects are still in the range of 20 to 40%. From this analysis we conclude that in the long run the effect on final energy use of an accelerated rate of energy efficiency improvement can be very substantial, even if such accelerated rate is only maintained for 15 years.

5.3
Renewable energy sources

Energy can be covered from a variety of sources. Box 1 gives an overview of the possibilities of the various sources. For a sustainable energy system, the application of renewable energy sources seems most appropriate.

Renewable energy already contributes 12–16% to the (commercial and non-commercial) global energy supply, mainly as traditional biomass and large-scale hydropower (see Appendix of chapter 4, fig. A.2). Various scenario studies indicate that in the second half of the next century half of the world energy demand can be covered from renewable sources (T.B. Johansson, H. Kelly, A.K.N. Reddy, R.H. Williams (eds.): Renewable Energy – Sources for Fuels and Electricity, Island Press, Washington, D.C., 1993; World Energy Assessment, UNDP/UNDESA/WEC, United Nations Development Programme, New York, 2000 (Chapter 7)).

In the following, different renewable energy options are discussed: biomass, wind and solar. In this section we focus on the options for Western Europe. A summary of all relevant energy resources is given in bx 5.1.

Biomass energy

Biomass is a generic term for all forms of energy derived from the biosphere (mainly plants), in a non-fossil form. Wood is the best-know example. At present wood is the most abundantly used renewable energy source and the most important source of energy for a large part of the world's population.

An important source of biomass is organic waste, e.g. domestic waste, manure, household and industrial wastewater, agricultural crop residues, residues from forestry. The estimates for the availability of residues in Western Europe range from 4–5 EJ (Pimm (2001)), which is somewhat less than 10% of present energy demand.

Next, biomass can be cultivated especially for energy purposes. Apart from wood, one can also think of a range of other crops, e.g. sugar cane, eucalyptus, sweet sorghum, miscanthus and sugar beet. For energy purposes, the productivity in terms of dry mass per hectare is important. For most biomass crops, where water availability is not a limiting factor, the productivity ranges from 10–30 t of dry matter per hectare; it depends on the type of crop, soil conditions and climate. Especially in arid regions productivities can be substantially lower, down to 2–4 t per hectare. For Europe, average yields are 10–12 t of dry matter per hectare. For the future, these values may increase, maybe to 15 t per hectare for the best soils.

Of course, the energy production depends on the area of available land. Table 5.4 gives an overview of land area in the European Union. In order to produce 6 EJ

Box 5.1 Possible energy sources for tomorrow's global energy system

The World Energy Assessment (Energy and the Challenge of Sustainability. World Energy Assessment, UNDP/OECD/WEC, 2000), which was published some years ago, gives a comprehensive overview of all energy resources presently known. It is very difficult, for various reasons, to predict the future potential of these options for the next 50 years. For instance, how would one quantify the potential of photovoltaic energy, considering the immense solar energy flux (see Chap. 2, Box 2)? The following considerations aim not so much at the theoretical potential of certain energy resources; they are rather based on an assessment whether a substantial share of the total demand of 600 to 1,000 EJ/year (20 to 25% of which as electricity) to be expected in 50 years can then be met by one particular resource, and what this would imply.

1. **Fossil energies:** Regarding the existing resources it would be quite possible, according to Tab. 4.3, that fossil energies still contribute the largest share to the global energy system in 2050. Yet this would entail the exploration of unconventional oil and gas deposits and/or a return to coal. Table 4.5 gives an overview of the consequences of this new strategy for the CO_2 system.
2. **Nuclear energy** (from fission and, in principle, from fusion too); *non-renewable*, but factually limitless with the technologies fully developed (reprocessing, breeders and, finally, fusion reactors), though only practical in a world that can be completely controlled politically. For instance, the present stock of power plants would have to grow 15- to 20-fold to make nuclear energy the main column of the future energy system. As explained in Chapter 4.2.4, the chances for this development are considered small.
3. **Hydropower;** *renewable*. Global capacity could be expanded to about 30 EJ/yr (today: 9.3 EJ/yr) in an economic way, and even to 50 EJ/J limited by technology only. However, the ecological consequences would be negative. Unsuitable as a significant contributor to meeting the growing energy demand of the future.
4. **Biomass,** *renewable*, existing capacity reserves (280 to 450 EJ/yr primary energy). If biomass were to play a significant role (>50%) in meeting the future energy demand, this would lead to an enormous demand for land, in competition with other kinds of land use. Already today, humans use 40% of the total primary production of planet earth to meet their needs (J.Swisher, D. Wilson: Renewable energy potentials, Energy 18 (1993) 437-459. D.O. Hall, F. Rosillo-Calle, R.H. Williams, J. Woods: Biomass for energy – supply prospects, in: Th. Johansson et al.,).
5. **Wind power;** *renewable,* has an enormous potential, in principle. If 4% of the sites suitable for wind farms were actually used, 230 EJ/yr of electricity could be generated.
6. **Photovoltaic;** *renewable*, virtually limitless potential in all regions (estimated between 1,500 and 50,000 EJ/yr of primary energy), but not economical yet at today's energy prices.
7. **Thermal and passive use of solar energy;** *renewable*, virtually limitless potential for energy demand at a low temperature level.
8. **Geothermal,** *conditionally renewable,* immense potential, but of limited use, from the present technical and economic perspective, where the geothermal flux at the surface is naturally strong (volcanoes, geysers).
9. **Ocean energy,** e.g. tidal and wave energy, *renewable*. Niche applications, no foreseeable significant role.

In a certain sense, **ambience heat,** which can be used through heat pumps, also belongs on this list. However, since it can only be used in combination with other energy carriers, it is classed as a method of improving the efficiency of the energy system.

(which is 10% of the present energy use in the EU), an area of 25 million hectare (250,000 km²) is needed[5]. This is one third of the area of arable land; 22% of the forest area or 12.5% of all the area potentially suitable for biomass production. Of course, there exists a substantial competition with other land uses. Some time ago expectations about setting aside land for energy production were high, but present prospects are uncertain[6].

Table 5.4 Overview of land area in the European Union in 1999 (million hectares). Source: FAO.

Forests and woodland	Permanent crops[7]	Arable land	Other	Total
113	11	75	115	314
36%	4%	24%	37%	100%

Biomass can be applied directly for energy purposes. Direct combustion is still the most common way of using biomass. It may range from small stoves in developing countries, to large industrial boilers, e.g. for the combustion of residues in the pulp and paper industry. In general, large-scale combustion delivers the most efficient and cleanest way of biomass utilization. For the time being it seems to be most interesting to apply co-firing of biomass in existing coal-fired power plants. Up to 20% of the coal in these power plants can be replaced by biomass; modification of the burners is required, but by now this is a proven technology. As in the EU the total coal combustion in power plants is 5.5 EJ (1999), this means that today an increase of biomass utilization does not depend on the introduction of a new conversion technology.

The future technology that is most discussed for biomass utilization is gasification. Under oxygen-limited conditions the biomass is converted to a mixture of gases like methane, hydrogen and carbon monoxide. This gas mixture can be used as energy carrier, e.g. for electricity production in highly efficient combined-cycle power plants. In addition, the gas mixture can be further converted to produce secondary fuels, like hydrogen, methanol or (via Fischer-Tropsch synthesis) synthetic gasoline. These fuels are suitable for automotive transportation.

For wet forms of biomass biological treatment processes are more appropriate. The best-known conversion process is anaerobic digestion. This is a bacterial process in a wet environment in which organic waste is converted to a mixture of methane and car-

---

[5] Assuming average production of 14 dry tons per hectare, a lower heating value of 18 GJ/ton and under consideration of the energy consumption for cultivation, harvest and processing.

[6] In the eighties in the European Community expectations about the amount of set-aside land were high. More than 15 million ha of land were expected to be taken out of farming by 2000 if surpluses and subsidies associated with the Common Agricultural Policy would be brought under control. However it turned out that these goals were not met. Reasons are both the restrictions in taking land out of agriculture and the limited demand for energy crops. The obligatory set-aside land for 2001/2002 amounts to about 4 million ha in addition to voluntary set-aside of 1.6 million ha i.e. in total 5.6 million ha. So far no new goals have been set on the amount of land to be set aside.

[7] E.g. fruit trees and vines.

bon dioxide. This process is especially suitable for wastewater and manure. Another process, fermentation, can be applied to, e.g., sugar cane and sugar beet to produce alcohol. However, for European conditions, this route to produce liquid fuels is considered less attractive (in terms of chain efficiency and costs) than the gasification-based routes described above. Under development – and considered very promising – is enzymatic hydrolysis that can be used to produce alcohol from wood.

Wind energy

Wind energy utilizes the kinetic energy in flowing air masses. In the eighties, wind turbines with typical capacities of 0.1 MW have been developed and installed. Since then a dramatic increase in unit capacity has taken place, with the biggest wind turbines now having a capacity of 2 MW. Until now most wind turbine capacity in the European Union is on land. By the end of the year 2001, total capacity was about 17,000 MW, with annual growth rates of 30%.

In wind farms it is customary to install between 5 and 10 MW of wind turbine capacity per square kilometer land surface. Various potential estimates have been made and optimism is increasing. Recent understanding is that for the European Union it seems possible to install 250,000 MW of wind turbine capacity on land and 150,000 offshore. The wind turbines would produce about 3 EJ of electricity. The 250,000 MW onshore would, if installed in wind parks, cover about 1% of Western European land, but the land could still be used for certain other purposes.

Solar energy

There are various forms of direct utilization of solar irradiation. Direct use of solar irradiation provides the highest energy production per unit area (see Table 5.5). The first form is heat production through solar collectors. Heat is irradiated on a surface that is thermally isolated from the environment. The heat can be carried away by, e.g., water or air. The most common utilization is for hot water production. Application for space heating becomes possible if the seasonal storage problem is solved. Typical conversion efficiencies from solar irradiation to moderate temperature heat are 30–60%. The total potential in the European Union is less than 1 EJ. An important limitation is the lack of (future) demand for low-temperature heat.

Much less developed, but more promising for the future, is direct conversion of solar energy through photovoltaic cells into electricity[8]. Efficiencies for practical systems reached by now are over 10%; but in the future efficiencies over 20% may be feasible. To date, photovoltaic (PV) power production is still among the most expensive renewable energy sources, but it is also the source that may become the most important in the long term.

---

[8] Electricity production through so-called 'solar thermal'. In this case, solar irradiation is concentrated; this makes it possible to generate high temperatures, e.g. hot air or steam. This is used to produce electricity. Concentration is only possible for direct irradiation; this limits the application to sunny regions, e.g. the south of Spain.

Table 5.5 Typical specific energy production per unit area from renewable resources.

Renewable energy technology	Energy production (MJ per m^2)	Energy form
Biomass	20–25	crude biomass
	10	electricity
Wind energy[9]	35–70	electricity
Solar collectors	1000–2000	heat
Solar photovoltaic	400	electricity

Table 5.6 Characteristics of global renewable energy use. For comparison: In 1998, total global energy use was about 380,000 PJ/year (380 EJ/year). Note that different energy forms are not fully comparable, e.g. 1 PJ of electricity from biomass replaces more fossil fuel than 1 PJ of heat from biomass. Also the economic value of 1 PJ electricity is higher than 1 PJ of heat. Source: World Energy Assessment.

Technology	Global energy production 1998 (PJ/year)	Increase in installed capacity in past five years (percent per year)	Current energy cost of new systems (¢/kWh)	Potential future energy costs (¢/kWh)
Biomass energy				
Electricity	580	~3	5–15	4–10
Heat	>2500	~3	1–5	1–5
Ethanol	420	~3	3–9	2–4
Wind electricity	65	~30	5–13	3–10
Solar photovoltaic electricity	2	~30	25–125	5–25
Solar thermal electricity	4	~5	12–18	4–10
Low-temperature solar heat	770	~8	3–20	2–10
Hydroelectricity				
Large	9000	~2	2–8	2–8
Small	320	~3	4–10	3–10
Geothermal energy				
Electricity	170	~4	2–10	1–8
Heat	150	~6	0.5–5	0.5–5

---

[9]   In order to produce 6 EJ (which is 10% of present energy use in the European Union) a production area of 25 million hectare is needed[23]. This is one third of the area of arable land, 22% of the forest area or 12.5% of all the area potentially suitable for biomass production.

Overview of renewable energy

An overview of the present use of renewable energy resources is given in Table 5.6. Many technologies are available already, but most of the so-called new renewable energy resources are expensive compared to conventional electricity production, twice as expensive for wind and biomass, and 10 times for photovoltaic solar energy. Technological learning leads to a reduction of costs per unit of energy produced (see figure 5.3), but this is a slow process.

5.4
Imagining the future: possible developments and effects

Taking into account the possible developments on both the demand and the supply side of the energy system, we present four different scenarios for the energy system in the EU in the year 2050: one with constant carbon dioxide emissions after 2010 and three contrasting cases, all of them involving a decrease in the level of carbon dioxide emissions by 75–85% compared to 1990 (and also a reduction in many other pollutants emissions). A business-as-usual type of scenario (indicated with '0') is added for comparison. It represents the continuation of existing trends, like small rates of energy efficiency improvement, gradually increasing final energy demand, an increasing share of natural gas, a phase-out of nuclear energy, and small shares of renewable energy sources. Note that the inclusion of this scenario should not suggest that for a period of fifty years such an image has more than an explorative character. Of course, this remark also applies to the other cases.

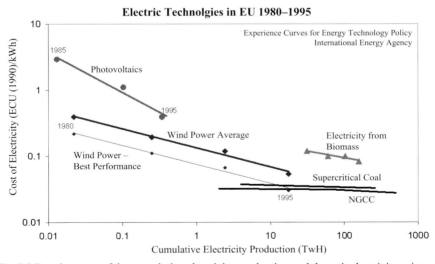

Fig. 5.3 Development of the cumulative electricity production and the unit electricity price of various electricity production technologies. Despite rapid learning rates, especially for photovoltaic, all renewable energy technologies still show higher electricity production costs than fossil-based technologies (supercritical coal fired power plants and NGCC: natural gas-fired combined-cycle power plants).

0. Reference case
I. A scenario with a fairly stable energy demand (however with a shift from heat demand to electricity demand) and a supply system that depends on the cheapest abundant sources available within the carbon dioxide constraint: biomass and natural gas.
II. Same development of demand, but with a smaller dependence on biomass.
III. A scenario with a strong reduction in energy demand.

The input of primary energy by fuel for each of the projections is given in figure 5.4. In scenario I it is assumed that low-temperature heat is supplied mainly by heat pumps and combined cycle district heating plants. Alternatively, a much more decentralized electricity production system is conceivable that is based on small-scale fuel cell generators that supply local heat demands. One third of the biomass production can be based on a variety of waste flows and residues. However, a substantial additional production of energy crops is required. The total land area needed to produce the biomass is 40 million hectares, which is 20% of the present area used for agriculture and forestry in the EU.

This vast amount of land requirement may be considered as problematic[10]. Instead, one may conceive a development that depends on a less space-intensive energy source, like photovoltaic solar energy. Development of this source requires a sustained growth of over 20% per year[11] in the first half of this century and a continuous investment in this energy source. This is presented in scenario II.

Another problem that may occur in scenario II is that the share of intermittent renewable energy resources, like wind energy and photovoltaic solar energy, becomes large. The degree to which intermittent renewables can be integrated into an electricity system strongly depends on the flexibility of the rest of the generating capacity, the availability of storage facilities and the way of power system control. If the non-intermittent generating capacity is sufficiently flexible, it seems possible to allow a share of 30–40% of wind and solar production without storage facilities, and of about 40–50% with sufficient storage facilities (K. Blok, E.A. Alsema, A.J.M. van Wijk, W.C. Turkenburg: The value of storage facilities in a renewable energy system, Proc. of the Sixth EC Photovoltaic Solar Energy Conference, Reidel, Dordrecht, 1985, p. 337–342). Therefore, the even higher share of intermittent renewables in the electricity sector in scenario II requires that part of the generated electricity has to be used for other purposes; a logical use of this energy is the generation of hydrogen through electrolysis for transportation purposes.

Stabilizing final energy demand, as assumed in scenarios I and II, already requires a substantial effort in addition to what may be expected to occur autonomously. Stabilizing energy demand can be considered to be the net effect of annual GDP growth of

---

[10] Note that alternatively biomass or biomass derived fuels can be imported from other continents. Yet, competition with other land-use claims will occur as well, like those for food, fodder and fibers.

[11] Note that a seemingly small increase of the annual growth rate of photovoltaic solar energy production capacity from 13 to 18% per year leads to a tremendous increase in the contribution of photovoltaic solar energy from 0.6 to 5.5 EJ_e. This illustrates our inability to judge on exponential growth for such long timeframes.

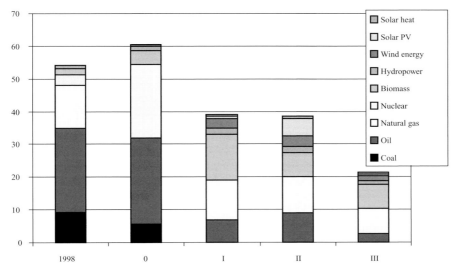

Fig. 5.4 Overview of the primary energy inputs in the year 2050 for the three images[12]. In the case of the wind, solar and hydro, primary energy is defined as the energy content of the electricity and heat produced. In the other cases primary energy is defined as the energy content of the fuel. This gives a suggested underestimation of the value of these renewable energy sources.

2–2½ %, structural effects leading to reduced energy intensity of about ½ % per year and an energy efficiency improvement of 1½–2 % per year (normally it amounts to 1% per year only). Nevertheless, as was shown in section 5.2, a substantial reduction in energy demand seems feasible, if technological progress in energy demand options is high enough. Scenario III shows a reduction of energy demand to about 40% compared to the present level. Compared to the other cases, all renewable inputs, except biomass which is the same as in case II, are reduced.

All three low-carbon images are characterized by a high growth rate of renewable energy, which indicates the transition that is required. In table 5.7, the necessary growth rates are listed. The difficulty of maintaining such growth rates over

---

[12] The basic data for the year 1998 are taken from IEA/OECD energy statistics. Final heat demand is calculated using present heat production efficiencies of 90% for industry and 80% for the other sectors. Non-energy use of energy carriers is left out of consideration.
   For images I and II it is assumed that total final demand is stable until 2050, but the share of electricity in final demand is assumed to rise from 20 to 27%. It is assumed that low-temperature heat demand is for a small part covered through solar heat, but mainly through electric heat pumps (coefficient-of-performance = 6). Industrial heat is covered by combined-generation-of-heat-and-power (electric efficiency 60%; heat efficiency = 30%). Heat not provided from renewable sources or CHP is assumed to be generated from natural gas (electric efficiency 70%). High-temperature industrial process heat is assumed to be produced from natural gas. Biomass is mainly used to produce automotive fuels (biomass-to-fuel conversion efficiency of 90%). The remainder of the fuel use, including those for petrochemical feedstock is from oil products. In image II, part of the fuel is produced from excess electricity through electrolysis (conversion efficiency 90%).

a period of 50 years cannot be ignored. Without any doubt, a continuous effort in invention/innovation and diffusion of new technologies is required to achieve such an ambitious goal. A summary of the different requirements is given in table 5.8.

Table 5.7 Required annual growth rates of the use of various energy sources in the low-carbon images I to III for the period 2000[13]–2050.

Energy source	I	II	III
Hydropower	1%	1%	0%
Biomass	4%	3%	2%
Wind energy	7%	7%	6%
Solar heat	11%	11%	11%
Solar photovoltaic	13%	18%	13%

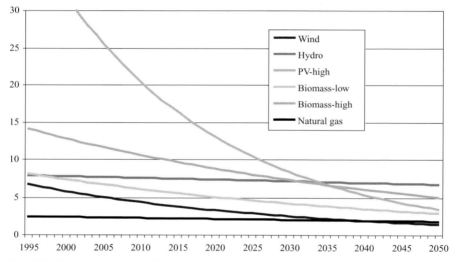

Fig. 5.5 Development over time of electricity production costs for various renewable resources assuming the growth needed for image I (growth rates are taken to be fixed). Costs are in dollar cents per kWh. The following progress rates are assumed (Energy Technology Price Trends and Learning, International Energy Agency, Paris, 1998): wind: 81%; photovoltaic: 82%; biomass: 70%; natural gas combined cycle: 90%. A progress rate of 90% means that costs are reduced by 10% for each doubling of cumulative production capacity.

---

[13] 1999 or 1998 if more recent data were not available.

Table 5.8 Overview of invention/innovation and diffusion necessary for the transition to a sustainable energy system.

Sector		Invention/innovation	Diffusion
Energy demand sectors	Industry	Development of a number of process innovations (see, e.g. Table 1) Uptake of more inventions to be developed to innovations Development of industrial processes Development of some cross-cutting technologies (high temperature heat pumps, heat exchangers, membranes)	Sufficient rate of implementation of new processes (e.g. through standards) Find methods to bring state-of-the-art technology to energy-extensive sectors Broad introduction of ambitious energy management systems
	Buildings	Development of better building shell components (also for existing buildings) Development of cost-effective heat pumps and fuel cells System approach to development of energy-efficient dwellings Development of energy-efficient electric appliances (best-available-technology approach for broad range of appliances)	Large-scale retrofitting of existing building stock Continuous sharpening of energy efficiency standards for new buildings and new appliance
	Transportation	Development of efficient light-weight cars with hybrid or fuel cell propulsion More focus on other efficient transport systems (efficient trucks, light rail)	Roll-out of new fuel infrastructure
Energy supply	Fossil energy use		Principle of no-heat-without-power to stimulate CHP Adoption of best-available-technology power plants
	Renewable energy	Development of advanced photovoltaic cells and integration in buildings and energy systems Development of advanced biomass conversion equipment (gasification, enzymatic hydrolysis) Development of off-shore wind energy converters Development of high-capacity heat storage systems	Long-term investment (partly non-profitable) in all renewable energy sources to enforce learning Set up a market and regulations for biomass energy feedstocks Arrange physical infrastructure and organization of electricity production in such a way that large-scale integration of small-scale production becomes feasible

Natural gas becomes the most important fossil fuel in all the scenarios presented. Nevertheless, the demand for natural gas (7.5–12 EJ) is somewhat lower than the actual demand in the European Union (13 EJ) and also lower than in a reference scenario. This demand would probably not cause a supply problem. First of all, Western Europe (including Norway) has still substantial reserves. Proven reserves are fairly small (about 200 EJ), but total conventional natural gas resources are estimated to be about 1000 EJ, and unconventional resources 1200 EJ (excluding the vast amount of methane hydrates (H.-H. Rogner: An Assessment of World Hydrocarbon Resources, Annual Review of Energy and Environment, 22(1997)217–262). Western European natural gas resources are 4–5% of world resources (see fig. 4.3). The largest natural gas resources are located in the former Soviet Union and in the Middle East. Increase of the import from Russia via pipelines and from other areas through liquefaction can add to the supply of natural gas to the European Union.

One important aspect is the development of the cost of the various secondary energy carriers. The costs for electricity production are estimated starting from the present production costs and using the idea of technological learning. A well-known rule of thumb is that costs of products decrease with a fixed fraction each time the cumulative production doubles (see chapter 2.3.2). The results depicted in Figure 5.5 demonstrate that learning is a fairly slow process and that it will take several decades before the costs of renewable resources drop to a level which is comparable with the cheapest conventional alternatives.

Finally, in table 5.9 a preliminary analysis is presented of each of the three scenarios with respect to the three aspects of sustainability. Although all scenarios show a strong reduction in greenhouse gas emissions, only scenario III satisfies the 2,000-Watt/cap criterion put forward in chapter 4. The other scenarios correspond to an energy use of about 3,600 Watt per capita, a value which is substantially lower than in a business-as-usual type of development represented by the zero projection.

German readers could ask, at this point, how the Energy Report 2001 of the *Bundesministerium für Wirtschaft und Technologie* must be judged in this context. In that report it is stated that a 40%-reduction of CO_2 emissions by 2020 leads to costs, which adversely affect economic growth.

The focus of the report on efficiency improvements in the buildings and transport area was too narrow. A significantly broader approach – as taken in this study, for instance – is needed to achieve the intended objective. The substantial efforts about regenerative energy technologies in German must be complemented by an equally strong engagement in the area of energy efficiency. Moreover, a much stronger engagement in research and development is necessary for ensuring a continuous development of new technologies. The expenditure in this field, which is low in Germany, at any rate, must grow considerably. This is called for in the present study.

Table 5.9 Preliminary comparison of the four future scenarios with respect to the three sustainability criteria.

Scenario	0	I	II	III
Energy use per capita	5,500 W	3,600 W	3,600 W	2,000 W
Economic	Costs of energy system will gradually decrease further. Risks of energy supply distortions.	Costs probably are somewhat higher than in image 0. High cost, including high transition costs.	Low investment image.	However, high upfront investments in RTD required.
		Substantial transition costs possible		
Social		Effect on European agriculture due to the shift to energy crops.		Substantial effort in energy efficiency improvement across all sectors required (probably small effect).
		Substantial structural changes, with associated employment effects. Total net effect on employment is hard to project, but some sectors, especially the coal and oil industry, will see substantial reductions.		
Ecological	Does not satisfy climate change criteria (carbon dioxide emissions higher than present levels). Other effects of energy production and use remain (e.g. brown coal mining, air quality effects) whereas others (acid deposition) maybe greatly reduced at a cost.	Most likely satisfy climate change criteria		
		Substantial land use requirements.		This scenario is, probably, most attractive from the environmental viewpoint (reduced energy and material flows).

5.5
Conclusions: What can be learned from history?

5.5.1
Sustainable energy technologies in the innovation trap

Concerning the question how far the (technical) efficiency potentials of energy supply and application are exploited, and to what extent new (regenerative) energy technologies are introduced in the market, one has to identify the specific conditions and obstacles faced by environmental technologies, which also comprise new energy technologies. Apart from the conditions that promote or hold back innovation, which are discussed in detail in Chapter 2, specific environmental technologies frequently find themselves in an "innovation trap" (more details in Steger, 1998).

First of all, it should be recognized that energy is not an important cost factor in most sectors. There are a few heavy industrial sectors where energy costs may reach up to about ten percent of the total production costs. However, for the vast majority of the industrial sectors, energy costs typically are about 1% of the total production costs or less (see figure 5.6). This is also true for the service and agricultural sector. For households energy expenditures typically amount to a few percent of their total expenditures. This means that for most decision makers, energy is not a very important factor when decisions on investments, purchasing and operational practice are taken. This causes the attention for energy issues as well as the interest for corresponding cost reduction measures to remain moderate.

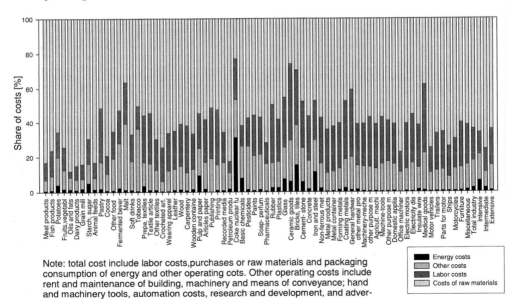

Note: total cost include labor costs, purchases or raw materials and packaging consumption of energy and other operating cots. Other operating costs include rent and maintenance of building, machinery and means of conveyance; hand and machinery tools, automation costs, research and development, and advertising expenses

- Energy costs
- Other costs
- Labor costs
- Costs of raw materials

Fig. 5.6 The share of energy costs and other cost components in the total costs of enterprises for various sectors in the Dutch economy. Source: data compiled by Andrea Ramirez, Utrecht University.

Secondly, if energy is considered as a cost factor at all, its positive potential in the market is underestimated because the positive external effects of measures (e.g. lower CO_2 emissions) are not, or not fully, taken into account, depending on the situation concerning (statutory) standards. More efficient energy technologies, for instance, often produce monetary cost savings too. But if energy is subsidized (as is the case e.g. in agriculture and in air transport), or if emission limits can also be kept to with technologies, the benefits of environmental sound and resource saving technologies do not take effect in the corporate investment calculation, which distorts the comparison of alternatives in favor of the conventional technologies.

Thirdly, new technologies are at the very "summit" of the learning curve. They are far less mature than the conventional technologies, which in some cases have undergone decades of a continuous improvement process. Small production quantities lead to (initially) high costs compared to conventional alternatives. The "4-stroke combustion engine" used in cars is a good example for the stability of one technology trajectory carried by mass production and continuous improvement. Although this type of propulsion energy involves specific drawbacks in terms of emissions and energy consumption, no potential alternative so far succeeded in touching the dominant position of the combustion engine in transport. It remains to be seen to what extent the fuel cell will bring actual changes beyond a niche existence in emission-limited areas e.g. large cities in California.

Being stuck at the upper end of the learning curve is particularly fatal for regenerative and decentralized energy technologies. As a rule, such technologies are "manufactured technologies", in contrast to the conventional "on-site-technologies" (e.g. power plants or refineries, which are erected in one site as a unique installation). Hence, the rapid realization of economics of scale is of strategic importance for the competitiveness of manufactured technologies against on-site technologies.

The fourth point is that energy technologies are rarely used as "stand-alone" systems. They are often embedded in grids or power lines. They must maintain compatibility within the chain by keeping to narrowly defined standards (e.g. petrol engines), or they are tailored to very specific applications as chosen by the buyers. Dedicated services and training centers have developed around the conventional technologies and their infrastructures. Such essential support structures are largely absent in the field of more efficient or regenerative energy technologies. Furthermore, the products of new technologies often do not fit into standards and complementary infrastructures, which have to be built up from scratch then

Fifth, the problem of "sunk cost" is particularly acute in the capital-intensive energy sector. Once a facility (e.g. a power plant or a refinery) has been erected, the capital raised for it has "sunk". Compared to early decommissioning (which entails high social and dismantling costs), any (continued) operation that still provides some collateral beyond the variable costs is economically rational. For the energy sector this means that new technologies have to compete with facilities that have been written off already (for instance nuclear power plants older than about 17 years, with a remaining service life of up to 35 years), or with installations that do not need full cost coverage. – The phenomenon of sunk costs also explains why unprofitable facilities are maintained for long periods in capital-intensive industries. – On the other hand, new investors may be deterred by the risk of getting involved in a price war.

Consequently, opportunities for new energy technologies lie in slowly growing markets, in the time window, which may be very narrow, when new investments are decided on, at the end of the economic lifespan of old facilities.

5.5.2
Substitution of energy carriers

The substitution of energy carriers is nothing new in the history of industrial development. While coal, as the energy carrier of the industrial revolution, first broke the limits of the previously dominant, regenerative energy carries and then replaced them, coal itself was increasingly replaced by oil since the beginning of the 20th century. This development did not occur at the same pace everywhere: In the US, where there were relatively few large coal mining regions, compared to Europe, and where the industrial structure of the, then, coal-based "tar chemistry" was not yet established to the same extent, oil could more easily conquer market shares as early as in the late 19-hundreds. In Europe, it was more difficult for oil to compete with coal. Only Churchill's strategic (i.e. not economic) decision to switch the British Royal Navy to oil in order to increase the range and speed of the vessels, and then the Great War brought the breakthrough of oil to secure market segments, most notably in the transport fuel sector. However, in regions where oil had to compete directly against coal, the substitution process, e.g. in navigation, stretched over decades. Before the Great War, there were only 500 oil-powered trade vessels. With the development of the Diesel engine and more effective oil-fueled boilers, the share increased to 54% by 1939 (24% for Diesel engines). After the World War II, which brought another push towards oil, this trend accelerated further: In 1957, only 8% of all trade vessels were powered by coal; by 1970 this share had shrunk to virtually zero.

The advantages of oil over coal in powering ships (and other transport systems) were obvious:

- Longer range with less bunker space (i.e. more freight space);
- Improved safety and easier operation with fewer staff;
- Quicker refueling.

The more the oil industry grew into a (or the first) global industry with falling production and transport costs, the more competitive became the price of oil. Today European coal can only be used (and can preserve market shares for the cheaper imported coal) in connection with state measures (electricity generation from coal, for instance, in Germany) or in specific technological applications (e.g. steel production).

However, after the oil price crisis of 1973/74, oil itself became the subject of substitution processes, especially substitution by natural gas in the areas of industrial and residential heating, and by nuclear energy in the electricity generation sector, where heavy oil power plants disappeared in many countries.

Considering the various substitutions of energy carriers, one can identify a number of common factors that were necessary for effecting a successful substitution:

- The new energy carrier must offer additional benefits beyond its economic advantages (e.g. oil is cleaner and easier to use than coal).

- Hardly any substitution was untouched by policy. Some political interventions were in favor of the new energy carrier, others favored the old carrier (often in the sequence: first support for the new energy carrier, then protection of the old ones, if the new competition turned out to be too successful).
- Once the new energy carrier has crossed the threshold of a "critical mass", its diffusion develops faster in its favor.
- The economic lifespan of the energy infrastructure that complements the fuel determines the speed of diffusion. (Large differences in the fuel costs or lower changeover costs can, of course, change the economic lifespan of facilities.)

These factors will also dominate the substitution of fossil energy carries by regenerative sources.

5.5.3
Final conclusions

The discussion so far has shown that energy efficiency innovations have a considerable potential to reduce, significantly, CO_2 emissions without giving rise to concerns about grave economic costs or structural ruptures. The reduction potential of these innovations is difficult to quantify for the long term, since we know little about the speed of the diffusion process by which the energy efficiency innovations will penetrate the market. This diffusion speed is not "given"; it rather depends on the actors, including state policy. As an important point for the recommendations for action, one must state, firstly, that accelerating the diffusion of energy efficiency can be an important lever for shaping a more sustainable energy system.

Secondly, it is clear now that the present targets for regenerative energy sources (e.g. generating 12% of electricity in Europe from regenerative sources by 2010) will not be met if they are not promoted massively in the future. (We will see that the "how" of this promotion is by no means a trivial question, even if the "whether" has been answered convincingly.) This applies, especially, to the long-term transition to a solar-based energy system.

Thirdly, the relationship between energy efficiency and regenerative energy sources has become clear: Even under optimistic assumptions concerning the development of regenerative energy sources, it is impossible to provide sufficient contributions to energy supplies, if the growth of energy consumption in the developed countries continues. Only if the energy efficiency potentials are exploited, i.e. if energy consumption is reduced, the regenerative sources can reach a share of ca. 50% in 2050.

As a fourth point, explorative considerations concerning earlier subsidies for energy carriers do suggest that the substitution is seen to develop faster once a "critical mass" has been achieved.

Again, the analysis of the "innovation trap" clearly shows that the innovation potentials of energy efficiency and regenerative sources will materialize quasi "automatically" or as a trend. Hence, there is an obvious need for action, which is not met to any sufficient extent, however, as the following analysis will show. We will examine the reasons for this before we develop our recommendations for action.

6 The Reality of Sustainability: Conflicts of Aims in the Choice of Instruments

The political discussion often tends to deal with the measures, i.e. the instruments to be employed, rather than the aims pursued with them. The ecotax discussion in several European countries is only one example among many. In such debates, any specific instrument either works miracles or leads to disaster, under the (model-type) assumptions made by its advocates or opponents, respectively. With regard to the objectives of the present study, such abstract or political discussion about instruments does not take us any further. In our analysis, an instrument can be assessed adequately only if (a) the aim for which it is to be employed is clearly defined, (b) the context is known (including e.g. the market conditions, other instruments already in use, cultural attitudes etc.) and (c) the distribution effects can also be estimated, since this is where one finds the grounds for opposing any given instrument.

Therefore, we would like to approach the issue from a more fundamental angle, firstly by making transparent the conflicts of aims that arise on the way towards sustainability. For demanding the integration of economic, ecological and social criteria as such does not imply the possibility of actually achieving such integration without encountering contradictions. Hence, we identify the relevant conflicts of aims in this chapter, from an "environmental aspect" – especially the reduction of energy-related CO_2 emissions – as our point of reference. The concrete analysis of instruments, where necessary, will follow in chapter 7 (recommendations for action).

6.1
Status of the theoretical discussion

The promotion of "welfare" or the "public weal" is often cited as the purpose of political or economic policy action. However, it is often unclear what this means when it comes to concrete measures of economic policy, because certain measures appear advantageous in some respect while, at the same time, they often cause drawbacks as well. An uneven distribution of the costs and benefits of a measure can be such a drawback; or, as sometimes is the case, the removal of one problem gives rise to a new one.

A well-known example is the magic quadrangle in macroeconomic policy: According to the law of stability and growth, a high level of employment, low inflation, external trade balance and appropriate growth must be the objectives. In fact, however, such a policy also aims at a fair distribution. These are five *aims of economic policy*, with the effect that measures for improving the achievement of one aim can easily compromise the realization of one or several other aims. If, for example, a higher level of employment is achieved, this can create a risk of higher inflation and more imports, and possibly leads to lower achievement with regard to the goals of "low inflation" and "trade balance". If one aims to achieve a fairer distribution of wage rises, this can lead to less employment and growth. Again, two aims may be met less successfully if one promotes one of the other objectives. Hence we are dealing with *conflicts of aims*.

The discussion surrounding such conflicts is pursued ever more meticulously, especially among economists (see Wagner 1989, for macroeconomic aspects, and Bhagwati and Srinivasan 1983, for the general theory of distorted balances). The effort is to find ways of improving the social situation for some individuals without putting others at a disadvantage. Such improvements are referred to as *pareto-improvements*; the resulting conditions – when no such improvement is possible anymore – are referred to as *pareto-optima* (section 3.2). To achieve such optima, and to identify deviations from them, one has to find social imperfections and appropriate, relevant measures. Another class of imperfections is referred to as *market imperfections*, i.e. the failure of the market mechanism. Measures for removing them and achieving the pareto-optimum are described as *economic policy instruments*.

Possible objectives are the reduction of market imperfections as well as certain distribution situations. Models in which market imperfections do not arise are also referred to as pareto-optimal models. Models in which x market imperfections occur are called "x-best equilibria", even in cases where policy instruments are employed in the best possible way. The optimal application of policy instruments looks for the best route between reducing one market imperfection and worsening another. Conflicts of aims even persist when one employs a number of policy instruments equal to the number of market imperfections. Where individuals have different preferences and attribute higher importance to e.g. environmental issues than other aims, they will want to put different values on the use of policy instruments and, hence, choose different positions concerning given trade-offs between aims. Conflicts of aims are thus accompanied by distribution problems on the benefit level because, for instance, somebody with one-sided "green" preferences gains more benefit if the environment is protected at the price of less employment, whereas someone with a strong preference for high employment achieves more benefits if employment is strengthened at the expense of the environmental aim. Since only *one* of the ideal solutions – ideal from the perspective of a given preference – can prevail, there remains only one of the many possible distributions of benefit, only one of the many possible allocations of the production factors, and hence only one of the many possible positions concerning the various trade-offs between the goals.

In the presence of conflicts of aims, the identification of appropriate measures suffers from the difficulty that one would have in knowing which of the imperfections that lead to conflicts of aims are more disturbing. In the above example, the question arises, for instance, what level of unemployment is so high that more environmental pollution should be sanctioned. However, even this will be judged differently by individuals with different preferences. What preferences eventually prevail is a question of political power. Each government – or the parliamentary majority in democratic systems – decides which aims are given a greater weight, and in many cases there will be a deadlock because a compromise that could carry the majority cannot be reached.

The purpose of this chapter is to elucidate the conflicts between environmental and other aims as far as these conflicts go back to market imperfections and distribution problems. Conflicts of aims, as long as they exist, can forestall political decisions, because individuals, especially politicians and lobbyists, can differ in their judgment of the importance of different aims. In particular, they can differ in whether they believe in the existence (or non-existence) of a problem or in their assessment of, or subjective judgment concerning a problem. It can become very expensive if no measures are taken because there is insufficient information even if a relevant problem is indeed very important, or if measures are taken which later turn out to be unnecessary. If, for example, there is no CO_2 policy pursued, although the costs of global warming can in fact be very high, greenhouse gases in the atmosphere will continue to accumulate, making expensive or quite impossible any belated policy towards a sustainable development, since the gases, once accumulated, cannot be removed anymore. Considering these possible costs of conflicts of aims, it appears sensible to reflect on the nature of such conflicts, and how they can be defused. This will be our effort in the following sections.

6.2
Environmental protection versus economic and social aims

Environmental emissions are regarded as market imperfections because the receiver of the emission has not signaled a demand for it through the market. On the contrary, in many cases he receives the emission against his wishes and would even be prepared to pay for reducing or removing the emission. One assumes, usually, that the polluter should pay for the pollution and that the pollution can be reduced in this way. A possible alternative is making the harmed party pay for the reduction. This idea will also be considered critically in the following because, especially concerning CO_2 emissions, almost everybody is a polluter.

6.2.1
Environment versus employment

Based on the polluter-pays principle, mandatory certification, a tax on CO_2 emissions or an energy tax are often suggested as instruments to reduce environmental emissions, in order to make the costs of the environmental strain a factor in the polluter's calculations. In this way, the polluter would have an incentive to prevent emissions. This beneficial effect, which is positive as such, has to be weighed against a probable negative effect on employment. For such taxes or certification costs increase the average and additional costs of a production unit. The price rise not only reduces domestic and foreign demand for the goods concerned, and hence their production and the emissions involved in it, but also the demand for labor and hence employment. This effect is all the more important if a corresponding price rise does not take place in other countries as well. For this reason, the EU precluded any unilateral action concerning a CO_2 or energy tax in the early 1990s. Opinion polls, again since the beginning of the 1990s, show that employment is regarded as more important than environmental protection (see Böhringer and Vogt 2001, p. 5, fig. 1). Schlegelmilch (2000) gives an account of the problems surrounding the for-

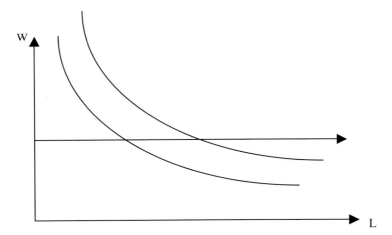

Fig. 6.1 Given a flat labor supply curve, lowering the taxes on labor costs while increasing the energy price shifts the decreasing labor demand curve towards the top right corner of the graph, to higher wages (w). A lower energy input has the opposite effect. If the first is the dominant effect, an ecotax can increase employment (L) in this case.

mation of majorities for the harmonization of energy taxes at a high level. According to his report, the problems arise because of concerns among governments with regard to competitiveness and employment.

The same result can also be derived from general equilibrium models.[1] Schneider (1997) showed this for a model assuming perfect competition, in which both higher pay and higher unemployment increase the efforts of the workforce. Ziesemer (2000) demonstrated the same for a model based on monopolistic competition, according to which unemployment is brought about by the searching costs for employers and employees that arise from any job vacancy. Hence, the conflict of aims is clearly the following: An environment tax or a solution based on certificates may reduce environmental emissions, eventually, but they may also give rise to higher unemployment.

In the debate about the so-called *double dividend*, a defusing of the conflict of aims has been proposed by using tax revenues for reducing the non-wage labor costs, which would, possibly, even increase employment (see Strand 1996, Schneider 1997, Bovenberg and van der Ploeg 1998, Koskela and Schöb 1999). If this were successful, one would have moved closer to two conflicting aims, simultaneously, even if neither of them would be fully achieved. Such models derive certain conditions under which unemployment would not rise. Empirically, however, it is still unclear – and, perhaps, will never be clear – whether these conditions are actually met. The underlying problem is that a lower energy input reduces the demand for labor, because it makes labor less productive. Higher energy prices, on the other

---

[1] The subsequent comments may only be accessible to professional economists. Readers less interested in theoretical and empirical details may like to move on to the last paragraph of chapter 6.2.1.

hand, increase the demand for labor, as a result of the substitution effect. For a flat labor supply curve with given taxes on labor costs, the only decisive condition is that the substitution effect outweighs the cost level and price effects; the result would then be fully determined by the demand change.

However, the labor supply curve does not have to be horizontal. According to empirical studies on the labor supply, the labor supply curve is rather an increasing function of the pay rate. The shape of this function depends on which theory of unemployment is applied. The relationship between the shifts of the labor supply and demand curves also depends on the national budget, since the reduction of one tax rate has to be paid for by increasing some other rate while the taxable base of both is changing as well. The efficiency wage model by Schneider (1997) predicts that, to keep their employees' effort high, employers themselves would raise wages as soon as employment increases and the tax rate decreases. This effect requires a high wage rate and, ultimately, increases (decreases) unemployment, provided the labor supply curve is sufficiently steep (flat) (see appendix 2). Only if the elasticity of wage rates is sufficiently low, and hence the correlation between taxes on wages and government spending negative, will unemployment fall with decreasing energy consumption (see Scholz 1998 and appendix 2). However, very little is known about the extent of such elasticities. If private employers are concerned that the elasticity of wage rates might not be sufficiently low, or the said correlation not negative, they will reject a tax reform that forces them either to set high wage rates and create unemployment or to increase emissions. Politicians will be worried that they are the ones who cause the increase in unemployment and will, therefore, rather oppose than support an environment-oriented tax reform.

Models in which wages are determined by negotiations between trade unions and employers rather than in a market of decentralized suppliers and buyers, on the other hand, produce a bargaining curve representing wages and unemployment ratios, instead of a labor supply curve. The position of the first depends on what incomes the unemployed and illegal employees can draw, and what negative benefits they derive from non-working hours. According to a model that includes "search unemployment", by Bovenberg and van der Ploeg (1998), an environmentally guided tax reform results in more employment only if the reduction of income taxes leads to an improved situation for the legally employed, but not for the unemployed and illegal employees (tax shifting). This shift of the tax burden only succeeds if, at the beginning, the incomes from illegal work or unemployment are sufficiently high, and the energy tax sufficiently low. The labor market remains unaffected if illegal employment does not exist and the energy tax is low, at the same time, or if the wages in illegal employment are not coupled to the market rates, because in this case the energy tax is a perfect substitute for income tax. If the energy tax is sufficiently high, the total tax burden rises and labor productivity falls (due to higher marginal costs), and this will dominate the shift effect. Using a more straightforward model not including search unemployment, Koskela and Schöb (1999) arrive at similar results as the ones just described. However, this is achieved, in part, by disregarding capital and energy as production factors.

Nielsen et al. (1995) consider the extreme case of a bargaining model where private employers have no power at all and where trade unions hold monopolist powers in every sector of industry and set a fixed wage rate, again leading to a horizontal supply curve. In a steady-state growth model of the Rebelo/Barro type, where government spending for education and emission reduction is productive, an optimum with five market imperfections is considered. With the preferences – which are affected negatively by environmental pollution – becoming "greener", the environment tax increases and the optimum growth settles at a lower rate. Consequently, the emissions also decrease, as well as the expenditure aimed at reducing them. Thus, taxes on labor can be reduced. The trade unions then set a lower wage rate, provided the unemployment benefits are not taxed as well. The result is a growing demand for labor and increased employment due to the higher demand in a situation of a horizontal supply curve.

The last three negotiation models allow the conclusion that an ecological tax reform will only succeed by diminishing the relative attraction of alternatives to employment, cutting bargained rents, and making wage settlements less advantageous for the trade unions. Hence, even the trade unions may then regard the price of a tax reform as too high. Including a sector of non-tradable goods in the bargaining model leads to additional elasticity problems (see appendix 2 on the model by Holmlund and Kolm, 2000).

Nevertheless, for bargaining models as well, there remains the critical question *whether* the returns rise or not. According to Koskela and Schöb, output only depends on labor input, which in turn is subject to decreasing returns to scale. The average product (output per unit of labor) exceeds the marginal product (output per *additional* unit of labor). The difference between the average product and the marginal product is a rent that falls to the business. If a "green tax reform" in the absence of market access leads to an increased labor input, the marginal product approaches the average product, the yields fall and the trade unions achieve only lower wages in the bargaining process, thus maintaining high employment. Therefore, in contrast to Schneider's efficiency wage model, there is no counterreaction along a wage or negotiation curve in this case. In the study by Bach et al. (2001), this becomes evident again in the partly empirical, partly simulative calculation of the effects of an ecotax. If the wages do not react to increased employment – once more resulting in a flat labor supply curve –, the ecotax has a positive employment effect. The higher the elasticity of wages in relation to employment, the smaller the employment effect, since the wages are pushed up by an ecotax. The possibility of a horizontal labor supply curve is justified by an estimate for the drop in the production of road vehicles, for which the unemployment rate was insignificant. The wage leadership of road vehicle manufacturing in the engineering industry is then used as a reason for applying the assumption of a horizontal labor supply curve to the entire economy. This justification appears to be insufficient, since wage leadership does not necessarily lead to identical labor supply curves. In the econometric, general (and therefore multi-sectorial) equilibrium model by Carraro et al. (1996), there are profits from imperfect competition. For the short term, a "green tax reform" increases employment and profits by lowering employees' income tax. The profit gain, however, makes trade unions negotiate higher wages, thus preventing any rise in employment. The wage-increasing counterreaction along the bargaining curve may cancel out the employment effect; this does indeed happen, according to estimates and simulations by the authors. Emissions are reduced for the short term but, in the long term, they rise back to the previous level. Still, according to their model, it would not help, either, to have weaker trade unions. The problem is the following: Because of an ecotax reform, strong trade unions achieve higher net wages, which contribute, through the income effect, to a higher demand for energy. Therefore, according to this model, emissions increase in the long term. If the trade unions are weaker, the wages lower and employment stronger, higher energy consumption in production leads to an even stronger increase in emissions than the higher private consumption that would be driven by higher wages. For the long term, their model predicts a positive correlation between employment and emissions. Only in the short-term, as long as the wage negotiations have not yet reacted, will employment strengthen and emissions decrease.

Obviously, the formulation of the labor market is of crucial importance here. This is confirmed by a review of 139 simulations taken from 56 studies (see Bosquet 2000). An improvement of the environment *and* of employment is achieved only under a number of conditions: (i) The rise in labor demand caused by higher energy prices has to outweigh the slump in labor demand caused by the reduction in energy input; (ii) revenues must be used for the reduction of labor costs, in order to ensure that labor demand will rise; (iii) the labor supply curve must not be too steep because, otherwise, the eventual increase in the labor demand will only push up wages; (iv) the substitution of energy and capital, which complement each other, by labor leads to a long-term fall in investment; this effect must be weak because, otherwise, there would be a long-term decline in employment; (v) the rise in employment, which could occur, must not be great because, if it were, emissions would increase.

Politicians who do not think in macroeconomic categories but rather in industrial sectors are concerned that the energy-intensive sectors would shrink while other sectors would be favored. This is expected to lead, initially, to redundancies in the energy-intensive sectors and only later, through market mechanisms, to new jobs. In other words: The direct effects of the policy destroy jobs and the indirect effects of market mechanisms create jobs. This will likely cause, at least temporarily, higher unemployment. At the present high level of unemployment in Europe, politicians are hardly in favor of running such risks. Therefore, we argue in chapter 7.2 that this conflict of aims can probably also be avoided through subsidies for energy-saving technologies. The direct effects would then lead to more employment while only the indirect ones would lead to a slump in employment. The latter is also referred to as a policy with positive incentives (Vermeend and van der Vaart 1997).

6.2.2
Environment versus the reduction of monopoly power

The conflict of aims between the environment and monopoly power has already been pointed out by Buchanan (1969). A monopolist wishing to maximize his profits sets a high price and the consumer demands, therefore, smaller than optimum quantities. If there were more suppliers, competition would be stronger, the price lower, and the consumer would buy larger quantities according to his demand. If one considers this market imperfection on its own, it becomes evident that economic policy should either aim at conditions involving many suppliers – for instance by organizing free market access, if necessary –, or it should regulate the monopolist in such a way that he limits the price he demands to the level of average costs. In both cases, consumers will receive higher quantities at a lower price than from the unbridled monopolist. The larger quantities can, however, also lead to more pollution of the environment. This actually happened in the energy sector itself at the end of the 1990s. Combined with the excess capacities already existing, deregulation (also see section 4.2.3 above) leads to lower electricity prices, a higher demand for electricity and thus, c.p., to more emissions as well (for the original literature on short- and long-term price elasticities, see Ziesemer 2000). This presents another conflict of aims. The competition policy should be less drastic in this case of a monopoly, because reducing the production volume through imperfect competition offers environmental advantages. Still, a closer analysis shows that the difficulty is to find the optimum volume that would represent the optimum middle course between these two problems. If, in the absence of any policy, the monopolist produces a smaller than optimal volume, the introduction of an environmental policy would even be harmful. If the monopolist produces a larger than optimal volume, in the absence of any policy, the introduction of an anti-monopoly policy is harmful while an environmental policy is beneficial. In this case, the deregulation of monopolies requires stronger measures to internalize environmental issues. The exact size of the optimum volume depends strongly on how the consumers of goods and environment judge, subjectively, the respective reductions in consumption or environmental quality. In a general equilibrium model, Soete and Ziesemer (1997) show how and why, in the absence of other economic policy instruments, the prices of environmental certificates differ from the optimal environment taxes. They assume (i) that competition policy does not intervene if, in a situation of monopolistic competition, the condition of zero profits is fulfilled and (ii) that, due to the variety of the goods supplied, externalities do not give cause for economic policy interventions. Then, the prices are at the level of the subjective assessment of the environmental conditions and do not take into account other market imperfections. Optimal environment taxes, on the other hand, do take into account other market imperfections. In the model, they can have negative effects if consumers attach great importance to goods and do not care much about the environment. The study by Radgen and Jochem (1999) also starts from the problem described here. Soete and Ziesemer (1997) consider monopolistic competition with *differentiated* products and free market access. The literature on homogeneous products and Cournot and Bertrand models assuming a given number of corporations are reviewed and generalized by Althammer and Buchholz (1999). In the literature about strategic trade policy, the aspect of profit-shifting, in particular, plays a role here. This aspect becomes important when dealing with sectors returning extraordinary profits. Whether these are of empirical relevance or not is still debated.

6.2.3
Environment versus trade liberalization

The advantages offered by international trade and its liberalization are based on the following:

– Consumers can purchase goods in the countries that produce at the most favorable cost.

- Because of this rise in demand, the countries concerned can specialize in such goods where they have cost advantages.
- This specialization leads to lower unit costs.
- One obtains other and more variants of goods from different countries than available domestically.

The first three arguments apply to the theories of both interindustry and intraindustry trade. For the environment, however, the expansion of international trade has the disadvantage that the longer international transport distances in connection with the trade also result in more pollution of the atmosphere, especially with greenhouse gases. Still, this should not lead to international trade restrictions but, in accordance with GATT/WTO rules and the theory behind them (see GATT 1992, Esty 1998), to a (internationally coordinated) environmental policy, which must also cover the international transport sector. If transport becomes more expensive because environmental costs have to be taken into account, the consumers of all countries will be more reluctant to buy goods that incur higher environmental and transport costs (see Soete and Ziesemer 1997); they will purchase more consumer goods domestically, and less abroad. Hence, an environmental policy not only reduces transport and pollution, as well as the decrease in gains from trade related to these factors, but also, as a side effect, the usual gains from trade that are made, as described above, as long as the pollution cannot be reduced without incurring higher costs. Consequently, it is difficult to convince the beneficiaries of these gains from trade of the advantages of environment taxes, certification systems and the inclusion of the transport sector in WTO-conforming concepts.

6.2.4
Environment versus capital movements

Another problem, the importance of which is controversial, concerns the induction of factor movements: If the same technology were used for every sector in all countries, one could expect, under additional conditions, that free international trade would align the factor prices in different countries. However, national environment taxes or certificates affecting various sectors to a different extent have the same effect as technological disparities. In such cases, a factor price adjustment cannot be expected anymore. In theory, if production factors are internationally mobile, capital will be moved into countries offering higher returns, while labor will migrate to countries offering higher wages. McGuire (1982) has shown that within a model based on three factors – labor, capital and use of the environment –, this process only stops when the more regulated sector has migrated abroad. If *pollution is a national occurrence that does not cross boundaries*, an improvement in the condition of the environment is achieved, however at a certain cost, because some sectors and, with them, labor demand have disappeared in the process (McGuire1982). Markusen et al. (1993) show that – according to models in which firms have a strong influence on prices and where the entry costs are too high to allow profits to evaporate in competition – the costs do not have to be too high if, inspite of environmental policies, the profits from domestic production (taking into account additional transport costs for exporting the goods) are higher than from producing abroad (including the fixed costs of setting up subsidiaries); therefore, sectors do not necessarily disappear. If, however, divisions of corporations go abroad because of high environmental prices, the costs will be all the higher.

If *pollution is a transboundary occurrence*, the only effect of environmental policy is to drive capital or divisions of corporations abroad. Pollution simply enters the country from across the border (McGuire 1982, Merrifield 1988). Empirical studies – which, by the way, do not focus on energy emissions, but on environmental costs in general – do not produce any evidence for pronounced capital movements as a reaction to environmental policies, because the costs are only a small fraction of the total costs and the same measures were introduced in the industrial countries at the same time (Cropper and Oates 1992). Thus, the effects on competitiveness may well be minor, primarily because competitiveness has always been an important consideration. As environmental policy might intervene more strongly, in the future, and the environment-related costs increase, capital flows could also swell. Furthermore, other policy fields apart from environmental policy can also demand cost-increasing taxes. Capital might begin to migrate as soon as many policies raise individual costs only slightly, but in total quite considerably. Environmental policies should not be judged merely quantitatively; they should also set an example for other economic policy decisions. For an empirically unexplored future, economic theory is well equipped to point out the risks that a more stringent environmental policy creates for other aims, as it has done above.

The issue of international competitiveness is closely connected with questions of capital movements. Several studies assert that environmental measures do have an influence, whose strength and relevance, however, depends on the authors' subjective judgment (Xu 1999, Letchumanan and Kodama 2000). Recent studies on the factor content of net exports show, in their majority, only small effects of environmental costs on unit costs. There is only one exception, which is Oceania (Xu und Song 2000), where the fraction of paid environmental costs in net exports rose sharply, compared to other Asian countries (Korea, Hong Kong, Singapore, Taiwan and the ASEAN countries). It is uncertain, yet, if this happened because environmental prices were established or because "dirty" industries migrated there, lured by a low but still positive price of the environment, and re-export from thence to the partner countries. The following results must be used with extreme caution, since the matrix of input/output coefficients valid for the US – including environmental factors – were applied to all countries. For "environmentally sensitive goods", the results reported by Xu (1999) even indicate that all countries, except for China, Japan and Norway, have *increased their comparative advantage* concerning the larger part of the trade volume in these goods.

The variable considered is the RCA (revealed comparative advantage) index. For the larger part of the trade volume, there is no change in the RCA with regard to the critical value 1. The percentage of the trade in these goods that shifted from non-specialization to specialization between 1965 and 1995 was: 32% in Belgium/Luxembourg, Brazil 46.8%, Indonesia 49.8%, Ireland 46.8%, Korea 42%, New Zealand 45.2%, Spain 40%, Taiwan 37.7%, UK 31.6%, and 67.7% in Venezuela. The increase in the environmental costs for the goods considered must be less than the increase in other cost components when compared to the cost increases of the trade partners. This means that significant disadvantage due to environmental policies can only be suspected for China, Japan and Norway. On the whole, the impression prevails that any change in the comparative advantages is dominated not by an increase, but by a decrease in the contribution from environmental costs in most countries.

6.2.5
Environment versus development policy

The tensions between environmental and development policies are constituted by two problem areas: the aftereffects of the "Clean Development Mechanism" (CDM) and the exchange of debts for environmental technologies, or rather the credibility of such environmental agreements.

The Clean Development Mechanism

The basis idea behind the CDM is that it is cheaper to achieve a given environmental effect in other countries than at home. This is, in principle, an efficiency gain. However, it can give rise to distribution effects that may conflict with development policies. If development aid is not a trade subsidy in disguise – i.e. if its purpose really is to fight poverty –, one has to ask the question what the CDM entails with regard to this struggle against poverty. One proposal among others in connection with the CDM is to run reforestation programs in developing countries (UNFCCC 2001, p. 9–11). If these increase the demand for land, they can lead to a rise in land prices. This, in turn, increases the costs for farmers who provide food for the local population. The logical consequence would be a rise in food prices. Poverty would become worse, putting this CDM measure into conflict with the aim of development policy, if the negative effects are dominant (concerning this point, also see Imboden 1993, p. 332). On the other hand, there is also the prospect of positive side effects such as reduced soil erosion. In the latest negotiations, limits were imposed for the possibility of CDM measures.

There is hardly any doubt that environmental problems in developing countries are often caused by poverty itself. But environmental policy can also increase poverty. We described in the previous paragraph how this can happen through a CDM measure. This should not be understood as an entirely negative representation of the CDM mechanism; but any judgment on it depends very much on whether it compensates or prevents deepening poverty among certain groups. If this is not the case, the evaluation of the CDM depends mainly on what weight one bestows on the increase in poverty.

Compensation for consumers and poor peasants could come from a land utilization tax, because landowners often have large land holdings at their disposal and would profit from a higher price for the factor land. These funds could be used for lowering any existing taxes on foodstuffs or for paying compensation to poor peasants, if the latter are prevented from shifting the increase in land rents onto food prices.

Transfer of clean technologies instead of environmental debts of sovereign states

Another problem with the inclusion of developing countries is of an institutional nature: Can countries that suffer from chronic difficulties with the enforcement of the rule of law and of tax payments guarantee that they will adhere to international environment agreements? Will they actually restrict their CO_2-emissions to the level permitted according to certificates purchased according to the international treaties? Considering such unresolved issues of control, sanctions and sovereignty (see Böhringer and Vogt 2001, p.8), we face the question whether it would not be better to steer the development of technologies in an energy-saving direction instead of striving for international agreements whose logical consistency has so far not even been formulated (see Böhringer and Vogt 2001, p. 3/4). Since developing countries import technologies from OECD countries, anyway, – and this will be the case for the foreseeable future, except for the countries that became or become members of the OECD themselves – technological development within the OECD will be one of the deciding factors for the energy demand in those countries too. An essential precondition for the reduction in technological dependence is the development of human capital that can master modern technologies. To date, even the poorer OECD countries have been unable to do this and suffer from severe comparative disadvantages in the engineering sector (SITC 7; also see chap. 9).

6.2.6
Environment versus supporting innovation

Technical progress in production processes is usually defined as a productivity rise of a given capital and labor input into such processes. Environmental economists often assume that the productivity of capital and labor is the higher the more environmental emissions are allowed (Pethig 1976). Under this assumption, a reduction of environmental emissions results in higher costs while technical progress leads, in principle, to cost reductions (McGuire 1982). Environment taxes or certificate costs forcing an increase in the marginal product of emissions is exactly the opposite result from the purpose pursued by promoting process innovations, i.e. an increase of the factor productivity.

One can think of two ways to escape from this dilemma. First, the increase in labor productivity could be stronger than the decrease in productivity caused by environmental regulation. Considering the fact that productivity growth has never exceeded 2% of the GDP per working hour over a prolonged period, this looks rather unlikely. Second, one could try to alter the direction of technical progress in such a way that the balance between saving labor and saving energy, the environment or emissions is tipped in favor of the latter (see e.g. Newell et al. 1999). An essential aspect in this is that one cannot assume, in the first place, that environmental costs are included in the calculations: Many environmentally sound technologies do not produce returns just by virtue of their environmental soundness. The market itself does not reward environmental soundness as such. Technology promotion, on the other hand, which does reward environmental aspects, could contribute to the internalization of positive or less negative effects just by offering positive incentives. Put differently, there are two essential externalities: negative external effects of production, transport and consumption, which cause environmental costs, and positive external effects of environmentally sound technologies, which avert environmental costs. The negative externalities require, in (first-best) principle, environment taxes or certificate solutions, which will meet wide-ranging resistance because they combine with other market imperfections in an unfavorable fashion. That is why they are unpopular politically. Whether this is justified depends on how environmental and other market imperfections are weighed against each other. The positive externalities, on the other hand, require subsidies to combine with the said market imperfections in a favorable way.

Still, there is the specific problem of "stranded assets" in connection with the accelerated introduction of new technologies compared to their "natural" diffusion into the market. As the *"Bauwerk Schweiz"* in chapter 4 illustrates, buildings and facilities have a certain service life, which is determined by technical and economic factors. The earlier investment decision was based on this period of use, although it is often more attractive, economically, to run facilities even beyond the point in time when they are written off, since after that, no more capital costs (write-down and interest on the capital committed) will arise. The dispute about the remaining period of use of German nuclear power plants demonstrated this in an exemplary manner. Accelerated innovation makes the old installations obsolete, economically, at an earlier stage so that they have to be scrapped. Depending on the structure and age of the capital stock, these "stranded assets" can be of considerable value and their early depreciation often leads to resistance against politically promoted innovations. This, of course, applies generally to the "process of creative destruction", as Schumpeter aptly described the innovation process, and has always led to efforts to delay, at least, this process through protectionist measures. Still, it is more difficult to organize resistance against market forces than against political decisions.

6.3
Standards arising from European law for weighing conflicting aims

Decision-makers try to resolve conflicts of aims by employing a variety of instruments. For, according to the "Tinbergen rule", the (unwelcome) side effects of an instrument increase with the "dose" at which they are applied; even if an instrument mix cannot cancel out the unintended (distribution) effects, they can still be alleviated considerably. A second approach is to "cast" rules for weighing aims into legal standards that define guidelines, criteria and methods for performing this weighing process. In the following, we will discuss considerations that arise from European law and are relevant to the present study as they must be taken into account for recommendations on strategies.

The European Community has long become an environmental community as well. The Amsterdam treaty strengthened this development considerably, especially by including, with Art. 6 EC, the environmental integration clause and, with Art. 2 EC, environmental protection as such in the fundamental provisions (for a detailed study, see Frenz/Unnerstall 1999, pp. 175–180). However, in the definition of duties in Art. 2 EC, environmental protection is quoted side by side with economic development and is connected to the latter in that this development has to be achieved in a sustainable manner. Thus, the conflict of aims between the economy and the environment is displayed as early as in the fundamental sections of the treaty. In the following, we are going to present some examples of issues of particular relevance for this study and demonstrate, in this way, which weighing processes according to which legal criteria lead to decisions on such a conflict of aims.

6.3.1
Free movement of goods

The economic development in the European economic region is based, essentially, on the achievement of the four fundamental economic freedoms: the free movement of goods, people, services and capital. In the energy sector, the free movement of goods has become the dominant issue. As energy is a cash-valued and tradable ware that is also standardized and intended for consumption, it has to be regarded as merchandise.[2] Through the concept of measures having an effect equal to volume restrictions on imports, Art. 28 EC disallows trade regulations imposed by member states in as far as they constitute a direct or indirect, factual or potential impediment to trade within the community.[3] The measure does not need to have a specific purpose in trade policy: If it affects or could hinder, objectively, the movement of goods,[4] it is prohibited. Thus, if state regulations actually have an obstructive effect on the movement of goods across boundaries, this too, is a relevant case. For instance, if electricity from other EU member states does not contribute to meeting a quota in favor of regenerative energies [in Germany] although it fulfills the national requirements in the supplier state, the foreign electricity suppliers affected are obstructed in their exporting to Germany. If, because of differing legislation in the country of origin, the imported electricity cannot be credited to the quota, the imports are failing to match the purpose of the German electricity supplier and will, therefore, not take place. This is not just a violation of the principle – which is integral to the free movement of goods – that any merchandise must be marketable in every member state if it meets the quality requirements in one member state.[5] Without any doubt, such discriminating treatment that does not specifically favor imported products, or at least treats them neutrally, constitutes an impediment to internal trade within the EU.

A quota system has the effect, generally, that producers from other EU member states, too, must look to this system if they wish to continue selling their goods in Germany. However, generally, Germany is not the main market for those suppliers. Hence, they have to make special efforts just for a "secondary market" – efforts that would not be necessary for other markets. The German producers, on the other hand, realize most of their sales in the national market. They too have to put in greater efforts, on the whole, but this burden as such is irrelevant, as restrictions imposed by community law do not have any leverage regarding national measures affecting domestic corporations. However, it is indeed relevant in terms of community law that the efforts of national corporations relate to a larger sales volume and thus can be undertaken more profitably by them than by suppliers from other EU member states. Ultimately, they can offer electricity at cheaper prices. In this way,

---

[2]   ECJ, Slg. 1994, I-1477 (1516) – Almelo.
[3]   ECJ, Slg. 1974, 837 (852) – Dassonville; Slg. 1995, I-1923 (1940) – Mars.
[4]   e.g. ECJ, Slg. 1978, 1935 (1954) – Eggers.
[5]   already in ECJ, Slg. 1979, 649 (662) – Rewe.

German companies gain a competitive advantage as a result of state regulation, which can lead to diminished market prospects and sales for other electricity producers (for more details, see Frenz 1997 and 2002).

6.3.2
Problems surrounding the EEG

§ 3 ff. EEG[6] lays down a purchasing and pricing commitment on the part of the grid operators in favor of the production of "regenerative electricity". All of the electricity offered by such facilities must be bought preferentially, and the electric power fed into the grid by them must be paid for at fixed minimum rates. However, according to § 2 EEG, only electricity produced in Germany benefits from this provision. The German electricity producers cannot meet part of their demand through purchases from suppliers resident in other EU member states, who are virtually excluded from exporting to Germany; the free movement of goods is thus obstructed. The ECJ, on the other hand, confirmed an environmental justification, pointing to the commitments of the Community and the member states arising from the Kyoto Protocol,[7] however without discussing further any alternative courses of action.[8]

6.3.3
Justification for restrictions for environmental reasons

If an increased production of electricity from regenerative energies is to contribute to cutting back CO_2 emissions, so that Germany is able to achieve the climate protection target according to her Kyoto commitment, or just to contribute to a general improvement of the air quality or the climate situation, then environmental protection is an appropriate ground for justification, which does not derive from Art. 30 EC, but has been firmly established, as an immanent barrier in extension of the Cassis jurisdiction,[9] since the ADBHU judgment of 1985.[10] However, regulation has to be non-discriminatory and proportionate, at least according to jurisdiction to date. Support for regenerative energies is discriminatory if it contains special provisions, different from those applying to German electricity, concerning electricity from abroad. One such difference is to stipulate regulations comparable to German law, since such a stipulation specifically disadvantages merchandise from other EU member states. In the view of the European Court of Justice (ECJ), discriminatory measures like this were justified only in cases, so far, where the discrimination could be attributed to requirements of environmental protection. Such a *requirement* was constituted by the principle of origin according to Art. 174 par. 2 s. 2 EC, which asserts that harm to the environment must be counteracted wherever it occurs.[11]

---

[6]   Parliamentary act on the preference for renewable energies (*Erneuerbare-Energien-Gesetz –* EEG) of 29 March 2000, BGBl. I p. 305.
[7]   ECJ, NVwZ 2001, 665 (666) – PreußenElektra.
[8]   hence crit. by Frenz 2002.
[9]   ECJ, Slg. 1979, 649 (662) – Rewe („Cassis de Dijon").
[10] ECJ, Slg. 1985, 531 (549) – ADBHU (*Association de défense des bruleurs d'huiles usagées*); Slg. 1988, 4607 (4630) – Danish returnable bottles; Slg. 1992, I-4431 (4480) – Walloon refuse.
[11] emphasized especially in ECJ, Slg. 1992, I-4431 (4480) – Walloon refuse.

Yet, this principle does not match the case of energy that is produced in a certain location before being fed into a grid and transferred, necessarily, to a buyer. Hence, in this case, the energy cannot be associated with one specific location. The ECJ, on the other hand, in its judgment on the *Stromeinspeisungsgesetz*[12] ["act on feeding electricity into the grid"], accepted a justification for discriminatory measures for reasons of environmental protection even where the previous condition – that the discrimination applies as a result of environmental protection requirements – is not fulfilled. This constitutes a considerable broadening of the possibilities for promoting domestic regenerative energies to the disadvantage of foreign electricity suppliers, at least if a (prohibited) subsidy is not imputed or even introduced.

In any case, a regulation restricting the movement of goods must be necessary.[13] The necessity can arise from the fact that Germany, in order to realize the objectives of the Kyoto Protocol, committed herself to reducing her contribution to CO_2 emissions by 21 % of the reference figure for 1990. Although the member states are due a degree of discretion, the proportionality of any measure is thoroughly scrutinized by the ECJ, which also demands substantiated, factually justifiable explanations in this case.[14]

As a first step, the Court examines the legitimacy of the objective. Environmental aspects must not be put up as a façade for purely economic considerations, which, on their own, cannot justify a restriction in the free movement of goods.[15] Hence, aspects of profitability do not suffice, even if environmental points of view are maintained formally.[16] They can play a role only if they are necessarily connected with objectives concerning environmental protection. Even then, though, there must be existing deficits in the area of the environment. At the same time, German industry has already pledged a self-commitment to climate protection by declaring voluntarily that it will make special efforts to reduce its CO_2 emissions or its energy consumption, respectively, by 28 % by the year 2005 and by 35 % by 2012, both on the basis of the 1990-figures. These pledges go beyond the German commitment in the context of the Kyoto Protocol. In this respect, there already is a movement towards achieving the objective – an activity whose success must not be endangered by additional forms of action that might involve other incentive effects. Considering the existence of considerable industry efforts to reduce CO_2 emissions, the need for government regulations appears questionable, especially for such regulations that already are the focus of scrutiny by the ECJ. State regulation is necessary only if there is no softer instrument to achieve the intended aim as effectively. The ECJ may not regard compulsory systems such as minimum quotas as unnecessary from the start, but it will examine them very carefully, especially in the context of the free movement of goods.

For instance, a waste reception system that makes it impossible for waste producers to pass refuse to other EU countries – possibly with the involvement of inter-

---

[12] ECJ, NVwZ 2001, 665 ff. – PreussenElektra.
[13] General comments on these requirements in ECJ, Slg. 1981, 1625 (1638); Slg. 1982, 3961.
[14] see ECJ, Slg. 1998, I-4075 (4127 f., 4132) – Düsseldorf.
[15] ECJ, Slg. 1998, I-1831 (1884) – Decker. Insofar, without closer inspection, also see ECJ, Slg. 1988, 4607 (4630 f.) – Danish returnable bottles.
[16] ECJ, Slg. 1998, I-4075 (4126 f.) – Düsseldorf.

mediate traders – was denied the status of being necessary.[17] National quota systems result in the opposite situation where imports from other EU countries are made more difficult or even impossible, if the barriers to market access are high enough. Setting up a deposit and return system for empties[18] is an admissible example, in principle, but it all depends on how it is done in detail, even in this case. A requirement, for instance, to use only packaging licensed by national authorities in such a system was judged to be disproportionate by the ECJ in cases where the authorities can deny the license even if the manufactures undertake to ensure the recycling of the packaging returned.[19] What becomes obvious here is that a voluntary measure taken by an industry can render government compulsion unnecessary. This view is supported by the concept of sustainability, whose intergenerational component shows that a sustainable development must look to the future. Therefore, the state must strive to achieve lasting results. Only then can the needs of future generations, whose ability to meet their own needs must not be compromised by today's actions, be taken into account adequately.

Energy policy plays a key role in this.[20] A lasting change in the behavior of economic subjects is most likely achieved if the subjects modify their behavior patterns reliably and of their own accord, i.e. out of conviction, even if only because any deviation from the new behavior would be punished with social contempt. Especially with regard to long-term objectives and a sustainable change of behavior, it is pointed out on the community level, and for the energy sector, in particular, that the industry should not just be addressed with government regulations, but should take part actively in solving the problems.[21] As a possible consequence, a dialogue with industry, voluntary agreements and other forms of self-control could be preferred, in the hope that regulatory intervention can be avoided in this way.[22] However, such voluntary measures have to be at least equally effective.

The EU Commission as well as various studies and a recent report by the Federal Environmental Agency in Germany regard self-commitments as operationally superior to regulatory instruments (Knebel et al. 1999). At the same time, however, the reasoning is differentiated and also emphasizes the superiority of certificates. At any rate, self-commitments must be seen to be successful. If that is the case, the self-commitments do not even have to be sufficiently binding, as the Commission has demanded. As far as they are agreed without any particular government pressure, they are, essentially, an expression of the self-shaping forces of industry. They defuse the conflict of aims between the environment and the economy, which itself is spanned by the principle of sustainable development.

---

[17] ECJ, Slg. 2000, p. I-3777 (3793 f.) – Sydhavnens Sten & Grus.
[18] ECJ, Slg. 1988, 4607 – Danish returnable bottles.
[19] ECJ, Slg. 1988, 4607 (4631 ff.) – Danish returnable bottles.
[20] already in the Resolution on a Community Program for Environmental Policy and Measures with regard to a Sustainable and Environmentally Sound Development, ABl. 1993 C 138, S. 31.
[21] Resolution 93/C/138/0 of the Council and the Representatives of the Governments of the Member States, meeting within the Council of 1 February 1993 on a Community Program for Environmental Policy and Measures with regard to a Sustainable and Environmentally Sound Development, ABl. Nr. C 138, Tz 11.
[22] already in the 5th Action Program of the Commission, KOM (92) 23 of 3 April 1992, Tz 31. More detailed in European Commission 1996 Tz 3 ff.; especially for the energy sector: *Frenz* 1999, 27 ff.

6.3.4
Aids and their justification

The Community-wide prohibition of subsidies (details in Frenz 1999) is part of European competition law. The exclusion of a certain group from a normative, universal system of obligations, such as the lightening of the energy tax burden on energy-intensive sectors, can also constitute a subsidy.[23] To a limited extent, the new Community framework for state environmental aids[24] may create more leeway in this respect. It considers all new tax reliefs and exemptions granted for ten years and representing a significant contribution to environmental protection (Tz. 50) as approvable for ten years and without degression, as long as an important part of the rates burden justified by them is paid for disregarding the lower national tax (Tz. 51 subpara. 1b indent 2). For a tax regulated by a Community guideline, the amount effectively paid must exceed the Community minimum (Tz. 51 sub-par. 1b indent 1). Such tax reliefs can be granted without these restrictions if, in return, the benefiting corporations commit themselves, in strictly controlled and sanction-enforced agreements, to achieving the intended environmental objectives or if they subject themselves to equally effective conditions (Tz. 51 sub-par. 1a). For existing taxes that are not changed, there must be a considerable positive effect on environmental protection, and the exception must be certain from the start or become necessary because of a significant degradation in the market conditions for the corporations concerned (Tz. 51 sub-par. 2) (For tax increases, the rules for new taxes apply, Tz. 52). To improve efficiency, conventional energy sources such as natural gas can also be supported (Tz. 51 Ziff. 3). For a national tax rate to be set below the rate agreed for the Community, an exceptional rule in the guideline is required (Tz. 49 lit. b). According to Tz. 54, aids for renewable energies are generally regarded as aids for environmental protection and are subject to less stringent conditions for three explicitly listed options: aids to compensate for the difference between production costs for renewable energies and market prices; support through certificates and bids; and access aids for new facilities on the basis of varied external costs. Any suspension of the prohibition of subsidies beyond this scope would also be in conflict with the polluter-pays principle, to which the 2001 Community framework for state environmental aid (Tz. 17 f.) recurs, in particular, where it is mentioned, fundamentally, as an argument against granting aids; the latter should only be considered as a temporary substitute solution, or as an incentive.

6.3.5
Possible ways of shaping the energy system following the ECJ judgment on the Stromeinspeisungsgesetz

Another way to avoid conflicts with the subsidies regime is to legislate with the intention of shaping the energy system. On 3 March 2001, the ECJ decided in the law suit PreussenElektra AG versus Schleswag AG that, according to the present status of Community law concerning the electricity market, the rules provided in

---

[23] general reference: ECJ, Slg. 1974, 709 (719).
[24] ABl. EC 2001, C 37, p. 3.

the *Stromeinspeisungsgesetz* ("act on feeding electricity into the grid") in its version of 24 April 1998 (StrEG 1998)[25], in force since 29 April 1998, where a purchasing and pricing commitment for electricity from renewable energies (RE-electricity) is laid down, do not violate Art. 28 EC. The aim of the StrEG 1998 is to reduce significantly the share of regenerative energy sources in the total volume of electricity production[26]. The StrEG 1998 has been repealed, in the meantime, and replaced by the act on the preference for renewable energies (*Erneuerbare-Energien-Gesetz*) of 29 March 2000 (EEG 2000)[27], in force since 1 April 2000. The principles, however, of the aforementioned rules of the earlier StrEG 1998, which were examined by the ECJ, remained unchanged in the new act. Hence, there arise no differences concerning the assessment in European law of the national rules on the commitment to purchase [renewable] electricity at a fixed minimum price[28].

In addition, the considerations of ECJ prepare the ground for other measures of an environment-oriented energy policy. In the RE-electricity judgment, the ECJ rejected the claim that one was dealing with a state subsidy as defined by Art. 87 f. EC, for the reason that the obligation of private electricity suppliers to buy electricity from renewable energy sources at fixed minimum prices above the actual [market] value of electricity and dividing the financial burden arising from the obligation to buy for the private electricity suppliers between the latter and the private operators of the primary electricity grids, did not lead to an indirect or direct transfer of state funds to the corporations producing this electricity[29]. The differentiation between "indirect" and "direct" serves the sole purpose of covering not only the advantages awarded directly by the state, but also such advantages that are granted via public or private institutions nominated by the state. Moreover, neither the fact that the obligation to purchase was based on an act of parliament and offered indisputable advantages to certain corporations, nor the fact that the financial burden resulting from the obligation to purchase had a negative effect on the business results of the corporations that are subject to this obligation and, thereby, could diminish the tax revenues of the state was sufficient to constitute a state subsidy.[30]

---

[25] Act on feeding electricity from renewable energies into the public grid (*Stromeinspeisungsgesetz*) of 7 December 1990, BGBl. I p. 2633; modified by Art. 5 of the act on securing the use of hard coal in electricity production and on altering the *Atomgesetz* and the *Stromeinspeisungsgesetzes* of 19 July 1994, BGBl. I p. 1618; last changed by Art. 3 of the *Gesetz zur Neuregelung des Energiewirtschaftsrechts* ("act on the revision of the law governing the energy industry") of 24 April 1998, BGBl. I p. 734.
[26] Salje, *Kommentar zum Stromeinspeisungsgesetz*, 1st edition 1999, § 1 Rn. 1.
[27] BGBl. I p. 305.
[28] For details on the alterations to the StrEG 1998 by the EEG 2000, see Büdenbender, *Die Entwicklung des Energierechts seit In-Kraft-Treten der Energierechtsreform von 1998*, DVBl. 2001, 952 ff.; also: Markard/Timpe, *Ist Ökostrom ein Auslaufmodell? – Die Auswirkungen des EEG auf den Markt für Grünen Strom*, ZfE 2000, 201/202 f.
[29] ECJ, NVwZ 2001, 665/666 – PreussenElektra AG/Schleswag AG with reference to national jurisdiction according to ECJ, Slg. 1978, 25/40 f. – Van Tiggele; Slg. 1993, I-887/933 f. – Sloman Neptun (second register); Slg. 1993, I-6185/6220 – Kirsammer-Hack; Slg. 1998, I-2629/2641 – Viscido a. o.; Slg. 1998, I-7907/7936 f. – Ecotrade; Slg. 1999, I-3735 – Piaggio.
[30] ECJ, NVwZ 2001, 665/666 – PreussenElektra AG/Schleswag Ag. For an extension to this, see Frenz, *Quoten, Zertifikate und Gemeinschaftsrecht*, DVBl. 2001, 673/681.

6.3.6
Competition and environmental protection

Self-commitments, like any collaboration between companies for improving the environmental quality of their products, can lead to conflicts between free competition as the basis of the free exchange of services between economic players, on the one hand, and environmental protection, on the other. Unilateral self-commitments on the part of industry as well as environment agreements can result in a variety of problems concerning competition law. Firstly, insofar as enterprises ["undertakings"] are involved, such collaborations themselves already constitute an agreement as defined in Art. 81 par. 1 EC. One is dealing with a resolution of an association of enterprises as defined in Art. 81 par. 1 EC when industry associations pass a resolution to enter a self-commitment or an agreement on environmental protection in accordance with their statutes.[31] Secondly, such instruments are often accompanied by cooperation between enterprises to achieve the objectives committed to in a combined effort; in terms of Art. 81 par. 1 EC, this constitutes, in any case, a "concerted practice". Companies that are not party to self-commitment measures are excluded from these common practices, making it far more difficult for such companies to meet the standards set in the self-commitment measures. As the consumers expect such standards, the excluded enterprises suffer competitive disadvantages causing, at least, potential damage to the trade between the member states and to a distortion of the competition within the Common Market.[32]

These effects occur also, and especially, when a self-commitment of all European enterprises (within a certain sector) excludes non-European suppliers. EU competition rules refer to all restrictions of competition with effects within the rules' geographical scope, independent of the origin of the author of such a restriction.[33] Hence, competition within the Common Market is shaped by suppliers from outside the EU. In the view of the Commission, EU-wide agreements could hamper, just "by their nature", trade between nations.[34] According to ECJ jurisdiction, this is also true for agreements applying to the entire territory of a member state, thus preventing the mutual market penetration that is among the objectives of the EC Treaty, and protecting domestic production.[35] The second constellation, in particular, will often emerge in the case of self-commitments.

When a state agency influences the achievement of self-committing agreements, this also constitutes a reason for the self-commitment itself, or the attitudes preceding or following the commitment, to impair competition. Hence the question arises

---

[31] A "decision" in terms of Art. 81 par. 1 EC is any statement of intent that is provided for, in statute, and arrived at in compliance with the rules of the statute, Commission, ABl. 1985 L 35, p. 20 (24) – fire insurance.

[32] On these two conditions, see e.g. ECJ, Slg. 1980, 2511 (2536 f.) – Lancome; Slg. 1980, 3775 (3791 f.) – L´Oreal.

[33] ECJ, Slg. 1971, 949 (959 f.) – Beguelin; not aimed at the effect principle, but at the place where the action is carried out: e.g. ECJ, Slg. 1972, 787 (838) – Ciba Geigy; Slg. 1973, 215 (242 f.) – Continental Can.

[34] Commission, ABl. 1983 L 376, S. 41 (47) – SABA II; ABl. 1985 L 19, S. 17 (21); ABl. 1985 L 20, S. 38 (42) – Ideal Standard.

[35] ECJ, Slg. 1975, 1491 (1515) – Belgian wallpaper manufacturers; also e.g. Slg. 1989, 2117 (2190) – Belasco.

whether the industry behavior instigated in this way may violate Art. 81 par. 1, 82 EC or the state behavior behind it gives rise to an examination of conformity with Community law. Articles 81 and 82 EC aim at preventing distortions of competition. Thus, it is important to establish actual causative contributions. On its own, state influence on the achievement of a voluntary self-commitment cannot exclude, generally, the relevance of industrial behavior, as far as competition law is concerned, unless it supersedes the causative effects of private behavior. Hypothetical causalities – the argument, for instance, that impediments to competition exist anyway, or that state regulation excludes any competition – cannot play a role, either (as already argued in Ehle 1996). The reference to the effects and the assessment of state measures is irrelevant also because of the fundamental difference between private and public law.

As a rule, unilateral self-commitments and environment agreements to reduce energy consumption are accompanied by innovations or other changes in the production processes. Thus, they contribute to the promotion of technical or economic progress pursuant to Art. 81 par. 3 EC. Since, by the provisions of Art. 2 EC, competition law is characterized as a general duty standard, the benefit to the environment constitutes such promotion. For the same reason, the reduction of environmental emissions is an appropriate benefit to the consumer, too, insofar as it exceeds the disadvantages resulting from competition restrictions.[36]

Since, as a rule, the cost increases caused by measures to reduce energy [consumption] should be small in comparison to the progress in reducing CO_2 emissions, and thus in climate protection, this criterion will be met. Generally, collaboration between enterprises in agreeing to, and fulfilling, commitments to reduce energy consumption is indispensable, because only in this way can any reduction be coordinated and any individual company or individual sector achieve objectives – by making use of synergy effects and common potentials – even beyond the individual enterprise and sector. Where the collaboration is limited to the energy sector, it will not be intensive enough, on the whole, to amount to the qualitative elimination of competition for a major part of the goods concerned, even if all enterprises of a certain sector cooperate at the European level. Thus, concerning unilateral self-commitments of enterprises and agreements on environmental issues, the conditions of Art. 81 par. 3 EC are fulfilled so that the stipulations of Art. 81 par. 1 EC can be declared inapplicable.

6.4
Energy-relevant research and technology policies of the European Union

Before we address recommendations for action in chapter 7, this section shall offer a brief overview of the present energy-relevant innovation policies at the European level (more details in appendix 3).

As shown in chapter 6.3, to integrate environmental policies towards a sustainable development with other EU policy areas has become a central demand in an increasing number of official EU documents. The environmental policy framework

---

[36] See Commission, ABl. 1971 L 10, S. 15 (22) – wall and floor tiles

is laid down in the 6th Environmental Action Program (KOM 2001, 31), which emphasizes four areas of action: climate protection, health and environment, nature and biodiversity, and the use of natural resources. Even if with varied emphasis, the energy sector represents a central policy area for solving problems in each of the other areas of action just cited.

The solution path has to keep to the canon of objectives of energy policy (economic soundness, security of procurement and environmental compatibility). To achieve these strategic aims, we need equally weighted initiatives for realizing an internal energy market, diversifying the energy sources and minimizing/internalizing the negative environmental effects of the transformation and und use of energy. The urgent demand for political action becomes obvious as soon as one considers our formerly monopolistic energy markets, which presently meet about 50% (with an upward tendency) of the demand through imports from politically sensitive regions, and the Kyoto Protocol with its CO_2 reduction requirements to be fulfilled by the signatory parties.

With regard to the timely development of new technologies and processes that save energy and increase energy efficiency, the increased use of renewable energy and new technologies in the field of engines and motors, the promotion of research, and technological development (RTD) are of strategic importance especially in this area.

The energy-relevant RTD programs of the European Union can be subdivided into five categories; in the following (and, in more detail, in appendix 3), we will focus on the first three of them:

1. The energy framework program (1998–2002) with its six specific programs
   - ALTENER (specific actions for greater penetration of renewable energy sources),
   - SAVE (specific actions for vigorous energy efficiency): energy efficiency/ saving,
   - ETAP (studies, analyses and prognoses concerning energy markets),
   - SYNERGY (international cooperation in the area of energy policies),
   - CARNOT (clean technologies for solid fuels),
   - SURE (safety, transport, cooperation in the area of nuclear energy);
2. The European Climate Change Program (ECCP; KOM (2000) 88)
3. The 6th Framework Program for Research, Technological Development and Demonstration Activities (2002–2006);
4. Third state programs: INCO, PHARE, TACIS;
5. Part of the structure program: e.g. INTERREG, RECHAR.

Framework program for activities in the energy sector (1998–2002), especially: ALTENER/SAVE

To achieve the energy policy goals cited earlier, Community initiatives for optimizing the transparency, coherence and coordination of all Community measures in the energy sector were formulated in the "framework program for activities in the energy sector" (1998–2002).

The programs ALTERNER and SAVE are not technology-orientated; they are intended to identify the legal, administrative and institutional impediments to an accelerated market penetration of existing innovative technologies and, ultimately, remove them by political means. In this way, ALTENER and SAVE represent a supplement to the technology-specific EU programs, where the SAVE program refers, approximately, to the energy demand side and the ALTERNER program more to the supply side. They start from where technology support programs usually loose their leverage, i.e. at the point of developing and evaluating activities to remove those impediments that hinder the market penetration of technically proven, clean and efficient technologies.

The European Climate Change Program (ECCP)

The ECCP was conceived to include every important interest group in the preparations for common, coordinated policies and measures to meet the emission reduction requirements arising from the Kyoto Protocol (KOM 2000, 88). The ECCP focuses on measures in the areas of comprehensive issues, energy, transport and industry. The proposed catalogue of measures takes into account, supports and supplements the efforts to integrate environmental issues with other policy areas. The ECCP also confirms the need to continue research in the areas of climate protection, technological development and innovation (KOM (2001) 580). For instance, it strongly recommends the more vigorous implementation and further development of the existing IPPC guidelines (Integrated Pollution Prevention and Control Directive 96/61/EC) and of the constantly updated state of the art technology in its reference documents concerned with IPPC energy-saving obligations. These measures can also be applied in the field of generic energy-saving technologies. Furthermore, it deals with issues of residential and industrial energy consumption (minimum efficiency requirements, energy demand management, promotion of nuclear power plants) as well as with a number of activities in accordance with the White Paper on a European Transport Policy (KOM (2001) 370).

The Sixth Framework Program for Research, Technological Development and Demonstration Activities (2002–2006)

In contrast to, and as a further development of, the 5th framework program (1998–2002), the basis and starting point for the 6th RTD framework program (2002–2006) is the concept of the European Research Area (ERA) (KOM (2000) 6).

In the present situation, there are "15 plus 1 research policies" – apart from those of the member states of the European Commission, which often act in parallel and with little coordination. The EC Treaty, on the other hand, provides in Art. 165 the express possibility that "Member States ... coordinate their research and technological development activities so as to ensure that national policies and Community policy are mutually consistent". However, this has hardly been put into practice yet (EVA 2001).

The new framework program is guided by the following basic principles:

1. Concentration on a limited number of priority research areas where a Union-wide approach offers most added value on the European level;
2. Developing the varied activities in view of their stronger structuring effects on research work in Europe, due to their close connection to the national, regional and other European initiatives;
3. Simplification and tightening of the procedural rules through re-defined forms of support and the planned decentralized administrative processes.

To achieve these goals, the EC framework program (total budget ca.16.270 bil. Euros) is structured around three main focus points:
– Grouping of research projects (ca. 13.285 bil. Euros),
– Shaping the European Research Area (ca. 2.655 bil. Euros),
– Strengthening the foundations of the European Research Area (ca. 330 mil. Euros).

The focus point: the "grouping of research projects", which brings together, and thus pre-structures, the research efforts and activities in seven priority research topics, seems to be of particular relevance. Energy-related research and development is to be afforded "appropriate priority". Hence, within the structure of topic areas, the sixth thematic priority, "sustainable development, global change and ecosystems" (budget: 2.120 bil. Euros) comprises the sub-programs "sustainable energy systems" (ca. 810 mil. Euros) and "sustainable surface and maritime transport" (ca. 610 mil. Euros). The key role of the energy system for a sustainable development, together with an advanced instrumentation, is taken account of in substance, but the funding is significantly reduced in comparison with the precursor programs: Compared to an "energy share" of 7.8 and 7.0%, respectively, of the total budget in the 4th and 5th framework program, the Joint Statement of the Council (2001/0053 (COD) of 30 January 2002) only provides for a share of 4.6 % (more details in appendix 3).

The short-term objectives of the 6th RTD framework program are:
– readying the improved technologies for renewable energies for the market and integrating them into grids and supply chains,
– promoting improvements in energy-saving and energy efficiency, primarily in towns and especially in buildings, and
– pressing ahead with the integration of alternative fuels into the transport system.

For the medium and long term, the central areas are:
– the further development of fuel cells, including their applications;
– the development of new technologies for energy carriers/distribution and energy storage, especially for hydrogen;
– new and advanced concepts for technologies in the area of renewable energies (primarily photovoltaic and biomass energies);
– collecting and bonding CO_2, as well as environmentally sounder installations for fossil fuels.

Concluding remarks

As defined in the Amsterdam Treaty, "the Community shall have as its task (…) to promote throughout the Community a harmonious, balanced and sustainable development of economic activities (…)". The equal consideration of economic, social and ecological interests, which should be aimed at in the context of a sustainable development and which is demanded in energy law, on the other hand, proves to present a partial conflict of aims, in which short-term economic interests dominate sustainable long-term requirements. The formal inclusion of sustainable elements into energy policies, for instance, is hampered by an inadequate financial background. The 6th RTD framework program outlined above may serve as an example of the imbalance between substantive and factual effectiveness. Because of inadequate funding, a program that points to the right direction, conceptionally, loses its effectiveness and, thus, precious time on a path towards a sustainable energy supply and production.

In this chapter, we mainly discussed the conflicts of aims between improving the environment and solving economic problems. We also discussed, in an exemplary manner, the legal considerations arising from such conflicts, and we presented an introduction to the energy-relevant RTD policies of the EU. In the next chapter, we will deal with the question as to how the conflicts of aims can be treated compatibly, and be avoided in some cases, by choosing the appropriate economic policy instruments.

7 Strategies for Accelerating Sustainable Energy Innovations

7.1
Reinstating energy as a strategic priority

Following the two oil crises of 1973/74 and 1979/80, there was a debate on the very principles of energy policy in all the western industrial countries. In those days, the question of "energy" was equivalent to asking: "How do we want to live in the future?" Nuclear energy was the example in the debate about the formative effects of energy technologies for societies and economies. Energy and environmental protection were also a common issue of conflicts. The "recycling" of the petrodollars was the big issue for financial markets in the process of globalization. The question of the extent of autarky concerning energy supplies led to the stabilization of coal in Germany and to large investments for the exploration of oil and gas sources outwith OPEC. The "oil shocks" triggered far-reaching changes – particularly as, and despite the fact that they were rather psychological in character and the price hikes which they induced receded again to some extent. A comparable political and economic priority for energy is absent today, apparently, although the risks have become more fundamental. The Kuwait crisis in 1991 was more or less ignored, globally. The dependence on oil from the Middle East has been strengthening again over recent years (see section 4.2) although, following the failure of the Oslo peace process, the region is more unstable than ever and the exhaustion of non-OPEC resources will eliminate the present "backlog" of production capacities – more likely leading to acute risks and escalating prices. It is highly probable that the development of populations and economies will catapult us not only past the existing production potentials of fossil energy resources, but also beyond the ecological limits (see chap. 4), while the hope for a solution of the energy problems through nuclear energy has withered away.

And yet, energy policy remains caught in the technocratic 'business as usual' approach. The EU Green Paper on Security of Supply – meant to be a "wake-up call" warning of the increasing risks surrounding the relevant imports – is still stewing in countless committee meetings, like other comparable documents, without any discernible consequence. The petrol price rises in Europe, in the summer of 2001, may have caused a lot of public excitement and indignation, but they rather served as a populist opportunity to introduce more energy subsidies (e.g. for farmers) and new protectionist demands (e.g. in the transport business). Nobody seemed to be interested in the link between oil price fluctuations and exchange rate volatility, nor in the concept of internalizing external effects, on which "eco-taxes" were based.

Its general use in all areas of life appears to make it difficult to direct prophylactic attention to energy. In the overwhelming majority of sectors, energy now contributes less than 1% to added value, and [residential] households have got used to the cost of energy: Inflation- and income-adjusted, the share of energy costs in private budgets is no higher than thirty years ago. The total expenditure on energy, on the other hand, has increased slightly, because of the strong rise in motorization (more than doubling the number of cars per 1,000 EU residents since 1970). Europe (and the US) seem to have lived with the risks surrounding oil from the Middle East for too long to cause any pressure of concern anymore. So far, every crisis that arose in previous decades was resolved, and whatever the regime is in any oil country, it will be keen to sell oil. This appears to be the only way of interpreting the present attitude. The climate issues are too far in the future, and their effects too uncertain to exert mobilizing pressure on energy consumption. Consequently, the EU is justifiably concerned as to whether the climate policy targets set for 2012 can actually be met (see KOM (2001) 226 fin.). Following this business-as-usual trend, the "energy carrier mix" in 20 to 30 years would still be similar to today's (which is dominated by fossil fuels, with an 8%-admixture from regenerative energy sources, compared to 6% today).

Possible synergies also receive little attention and are thus not being used. The Integrated Pollution Prevention and Control (IPPC) Directive of the EU, for instance, is based on the assumption that, in industries that put strains on the environment, there has to be a continuous trend (or, more precisely: pressure) towards using the 'best available technologies' (BAT), which are specified in voluminous regulatory documents. This directive could be easily applied to the energy-intensive sector as well, if sufficient attention were paid to this aspect of environmental pollution, and if administrations were to put the directive into action consistently.

In this situation, it is *the central task of political leadership* to put energy back on the agenda as a strategic priority for the economy, citizens and politics. This task has two decisive dimensions, one in content (chap. 7) and one in procedure (chap. 8). The first dimension is defined by the question as to *what* we shall do, the second by *how* to implement our decision, i.e. how to overcome the conflicts of goals discussed in chapter 6, and then how to realize our goal operatively in all the details (which is not debated here any further, because the present study is oriented to the more distant future, and has to be left to separate specialized studies).

What to do depends on analysis and objectives. The result recorded in chapter 4 – which was obtained through the criteria of sustainability developed earlier – was that the present level of energy consumption and its composition (contribution from fossil fuels, and from the Middle East) is unsustainable. A "2,000-watt benchmark", which would be more in line with criteria of sustainability, especially if one considers the long-term population and economic development of populations on a global scale, was offered for guidance. Other studies arrive at similar results, most notably if they are based on the "moderate" IPCC scenario S450, as is the European Climate Change Programme, for instance. In practice, this would mean an annual reduction of the CO_2 intensity by 4% (see section 4.3.1). A rough estimate using the concept of a "time of safe practice" shows that there seems to be enough time left (as yet) for a long-term reorganization of the energy system, although this task should be attacked with urgency. For the existing potentials do not turn into reality simply by themselves, as was shown in chapter 5. Not only technologies, but also markets must be created *now* so that these potentials can be exploited *in the future*.

To this end, however, long-term goals must be formulated in a clear and plausible fashion. Presently, the policies are still strongly fixated on the "Kyoto period" of 2008–2012. This is necessary, but not sufficient. The goals to achieve by 2020 and 2050 must be formulated and communicated more clearly now, especially as it takes a long time to arrive at a consensus on the European and international level. In this, it appears necessary to "convert" those goals into plausible quantities, which have to be verified continuously (e.g. the proposed reduction of the CO_2 intensity by 4% p.a.).

The second strategic lever – apart from the goals – is the focus on innovations. Especially because of the numerous political and administrative obstacles, is this lever, which is regarded as relatively "painless", so important. While the alternative allocation of resources can easily be viewed as a "zero sum game" that involves tough distribution battles, as a consequence, innovations facilitate new strategies for compromises, because the losses that may ensue are only "relative". The reason for this is that innovations, per definitionem, offer better value for money than the solutions replaced by them (see section 2.3), i.e. they are somewhere in the spectrum between the two "extremes" of a higher energy efficiency and thus reduced emissions and an unchanged energy efficiency at lower costs. The competitiveness of photovoltaic electricity (see chap. 5) can only be improved through innovations resulting in higher efficiencies. Based on such innovations, those "economies of scale", that are the very reason for the competitive advantage of decentralized, "manufactured" technologies over conventional production units erected on site, can be achieved in the course of market introduction and diffusion (see section 7.3).

In practice, however, the priority given to innovations cannot be assumed – beyond rhetoric –in general, as the example of the "campaign for take-off" shows exemplarily. With this campaign, the EU Commission aimed at advancing capacities for 10 GW, each, from wind energy and biomass and 1 GW from photovoltaic electricity, as well as 1.5 GW of power in solitary settlements (communities aiming, therefore, at covering all their energy demands from renewable sources). Instead of producing a badly needed 'technology push', the project did not get over the financing hurdle (that is, other tasks than supporting innovations in the field of renewable energies prevailed in the political process). Hence, the objective of the EU Commission to increase the contribution from regenerative energy sources from 6% to 12% (electricity: from 14% to 22%) by the year 2010 is now in danger, which also renders the goal of producing about 50% of the 2,000-watt benchmark from regenerative energy sources by 2050 less realistic, since the development and diffusion of technologies cannot be accelerated *at will*.

These are the perspectives from which to consider the recommendations for action, which will be developed in the following. Even if there is not always a clear distinction between the various fields of action, or the different phases of innovation virtually melt into each other, it is still preferable, for the purpose of explanation, to differentiate between them in the way it is done here.

In keeping with the objective of our study, we focus on energy innovations here, however, fully aware of the fact that the overall effect can only be achieved through wider policies towards sustainability, which will cause necessary changes in many areas. There are three strategic fields of action concerning energy innovations, at which the following measures are directed:

- The framework conditions, on the whole, must be made more supportive of energy innovations.
- The supply and demand conditions for innovations offering higher energy efficiency must be shaped in such a way that the potential derived in chapter 5 can prevail earlier than under status-quo conditions.
- The regenerative energy sources must be developed, and made marketable, more rapidly; the structural shift from fossil to regenerative energy sources needs to be accelerated.

Generally, the technological innovation potential is to be fully exploited to meet the "2,000-watt benchmark". This requires an instrument mix which is tailored to the individual phases of the innovation cycle and takes into account each specific area of application: homes, industry and transport.

Due to the severe uncertainties and the multitude of factors influencing the innovation process, we are not in a position to quantify the effects of the measures proposed to this end. In trying this, we would only repeat the strategic mistakes made in earlier energy forecasts. The crucial point is – as justified in detail in chapter 2 – that the learning, experimental and realization process concerning sustainable energy innovations is accelerated.

7.2
Improving the framework conditions

7.2.1
Defining the limits of using natural resources

Innovations require visions or, at least, challenges. Space research – the idea of having a man landing on the moon, in particular – is an example of how certain visions can drive and inspire the innovation process over a long period. Before climate protection and the scarcity of fossil resources can be seen as a challenge to all actors in a society, these scarcities must first be made visible through the definitive formulation of limits to the use such resources. As soon as these "crash barriers" for social and economic development have been erected, the innovation process will also take a new direction. Visions as, for instance, the "era of solar hydrogen" can emerge, then, in response to the need for reducing the use of fossil energy resources.

Therefore, formulating guideline quantities such as the "2,000-watt benchmark" is of fundamental importance for sustainable innovations. They constitute a plausible starting point – in scientific terms – for the social discussion about the structure of the energy system. The direction of innovation activity will change as soon as new priorities crystallize in this process and definitive goals begin to emerge. Such changes in the framework conditions affect the expectations formed by businesses. The latter also react, in their strategic planning, to "soft" signals, long before these are actually reflected in market prices. Innovative enterprises, in particular, operate proactively; they do not act only when new chance-risk constellations are obvious to all players. Energy efficiency solutions and substitutes for fossil energy carriers are recognized as new fields for "extraordinary profits". In this respect, (ecological) restrictions represent new opportunities as well.

To change the direction of the innovation process, it is sufficient, initially, to attract an innovative elite. As soon as this elite is reassured by initial successes – helped by improved framework conditions – the mass of imitators will follow them in the new direction. Hence, a long-term innovation policy must, above all, take care of the pioneers who have the capacity to set new paradigms. A suggestion for how to activate these pioneers or elites will be offered in chapter 8.

7.2.2
Using the market: Signs of scarcities induce sustainable innovation

The role of the state is not limited to moderating and institutionalizing the process of formulating goals. Government action is required, above all, in translating the soft signals of the goal formulation process into hard market signals. This can be done in various ways, always keeping in mind the conflicts of goals discussed in the previous chapter.

The most obvious and effective approach appears to be starting from the remaining usable volumes (e.g. the GGE or CO_2 emissions regarded as allowable). These can be allocated as property rights to the users/polluters, following various procedures (free primary allocation to current users – "grandfathering" – or auction). The gradual depreciation of these property rights (certificates, licenses) with concurrent economic growth creates an innovation pressure. Corporations that do not respond in an innovative fashion pay a high (certificate) price, which in turn acts as a monetary incentive to innovate. Still, there are circumstances where the certificate model is not practicable, for instance there is a large number of diffuse sources or polluters (CO_2 from mobility). Aside from that, establishing the remaining usable volumes is anything but a trivial task.

Hence, tax solutions ("eco-taxes"), which ensure a reduction of environmental pollution (CO_2) by laying down a price (increase) path, are an alternative to be considered. While being ecologically less effective, eco-taxes offer the advantage of additional public revenue, which can be used for reducing other (efficiency-decreasing) taxes or duties. Their function as an innovation incentive remains intact. Any innovation that saves a pollution unit presents a cost relief (through a lower eco-tax bill). Combinations of certificates (for major sources of pollution) and tax (e.g. on fossil fuels) solutions are also conceivable. Yet, the effects strongly depend on the specific conditions and on how such solutions are structured (see section 6.2), and there is considerable resistance.

In contrast to such market solutions, a regulatory approach yields rather modest long-term innovation effects. In the short term, the accelerating effect resulting from the prescribed introduction of a new technology appears attractive (as the 'best available technology' is declared to be the 'state of the art'). Experience tells us, however, that this accelerating effect is offset by a retarding effect resulting from discrimination against alternative solutions (e.g. fuel technology against flue gas desulphurization), on the one hand, and from the tendency of (outdated) technical standards to prevail. A suitable field of application might be e.g. prescriptions for the stand-by losses of household appliances, where various efficient technical solutions already exist, whose market introduction would be ensured in this way.

Ultimately, whatever instruments are employed, the decisive factor in the long term is the adjustment of price ratios (see section 2.3.2, sub-par. 2). Changes in price ratios refocus the search efforts in the infinite pool of inventions and ideas. With rising energy prices, the interest in, and the chances of realizing energy-intensive projects will dwindle. Investors will turn their attention to energy efficiency solutions possibilities of substitution, but only under the condition that the change in the price ratios is perceived as permanent and not as a mere transitory fluctuation. To meet this condition, sustainability as a goal, and the instruments chosen for achieving it, must be based on a broad (independent of any change of government) and robust (unaffected by short-term negative effects) consensus. Price ratio changes and innovations thus induced will lead to an ecological structural change and will hence result in the annihilation of market positions and assets. For companies, adjustment issues will arise especially from existing technologies that cannot be changed quickly; for private households, such problems are caused by inflexible, traditional consumer attitudes. To prevent a considerable depreciation of capital, beyond the usual investment cycle, and to avoid negative distribution effects, continuity must be ensured when reducing the CO_2 emissions.

As part of an adjustment of price ratios, subsidy policies must also be inspected, especially concerning such subsidies that favor non-sustainable developments (also see SRU 1996, p. 331, 398 f.). Even today, energy consumption is still subsidized in many ways, either directly (e.g. air traffic) or indirectly (e.g. through residential heating subsidies). Where subsidies for energy carriers are granted without any checks on energy efficiency and environmental effects, the removal of such grants generates financial resources for supporting sustainable energy innovations. However, the price-induced refocusing of the search efforts will have consequences only if interesting projects with a potential towards sustainability actually exist in the pool of inventions and ideas. In this respect, changed price ratios are not a sufficient condition for changing the direction of innovation. This leaves us with the question as to how the pool of inventions and ideas can be influenced, which will be treated in the following section.

7.2.3
Providing infrastructures and generating competences towards sustainability (the technology push)

Long-term innovation policies must also wield influence concerning the main topics of research and the competences imparted through the education system. These (infra)structural prerequisites of innovation activity change only slowly; yet they are a crucial determinant for what can actually be done in the innovation process. As a supply-side policy, the publicly financed generation of capacities (in human capital, knowledge capital and the associated "hardware", such as education and research institutes and communication networks) can play an important role in setting the course at the pre-competitive stage. A certain supply of ideas and competencies will create new solutions and markets. For instance, a nation or region that offers an ample supply of ITC/EDP/computer science specialists will experience an innovation push in this sector (predominantly through new entrepreneurs). This is equally true for the biotechnology sector (for empirical evidence, see Staudt et al. 2001). Thus, state input in the area of basic research and education policy is of similar importance for sustainable development as changes in price ratios.

If these considerations are correct, the next step would be to ask what reforms of research and education policy are necessary and most urgently needed to prepare for sustainable energy innovations in particular.

— *Research policy:* Due to its strategic importance, funds for fundamental research in the (non-nuclear) energy sector should be increased significantly (also see SRU 2000, p. 541). Doubling these funds within the coming 4 years would be an appropriate start signal. Public research funds for projects in e.g. industrial process technology and materials research, among whose objectives is a significant improvement in energy efficiency, should also be strengthened.

— *Education policy:* Which competences are important for sustainable development? Which disciplines are most challenged to change their course contents? The innovation process in the energy sector certainly requires a stronger awareness of energy-efficiency issues in the training of managers and engineers. Unused potentials also exist in the training of craftsmen and architects. Imparting the relevant problem awareness and problem solving competence in these fields has far-reaching effects. However, with the increasing autonomy of higher education institutions, the options for the state to take immediate, formative action are diminishing. Third-party research grants, on the other hand, which also radiate into teaching, become more important.

Investment in the forms of national wealth mentioned here are essential not least because future generations would be left not only with exhausted natural resources, but also with an "exhausted" knowledge base, if such investment does not take place. In addition to this, investment in modernizing the (hardware) infrastructure of the national economy is necessary, e.g. in an efficient transport system. Apart from the transport route network, this also includes electronic traffic control systems (in connection with economic incentives, e.g. "road pricing") or attractive regional public transport systems (also see section 7.7).

7.3
Action field energy efficiency in industry: Accelerated market introduction through subsidies

In chapter 6, we discussed the conflicts of goals that can arise between the different goal dimensions of sustainability (economic, ecologic and social) and hence between the corresponding fields of policy. Considering these tensions, (eco-)taxes are just one possible instrument for bringing about the change in price ratios that is necessary for a sustainable innovation policy. Positive incentives should be used increasingly, too. Therefore, one finds increasing preparedness on the political level: a will to engage in a policy of positive incentives instead of relying on the polluter-pays principle, which is equivalent to a burden on industry (KOM (2001) 68 fin.).

In the following, the effects of a policy of incentives through subsidies for energy saving innovations are examined, using a Dutch model as a reference point for an

EU-wide recommendation. We are mainly dealing with energy-efficient technologies that are in the phase of market introduction, meaning that pilot and demonstration projects are already in place and the task is now to attract an "early adapter" and develop industrial production and service structures with an increased number of units.

These subsidies mostly represent start-up financing, since one can show (Steger 1998, p. 216), the costs of the cleaner technology do not exceed the costs of the old technology in the majority of cases. This finding is illustrated in fig. 7.1. The typical cost developments are shown here:

- The costs of the new technology are always higher because it incurs higher capital and operating costs (curve A1). Therefore, it will be used – if at all – only in very specific circumstances (electrically powered cars in health resorts), and will not play any role as a standard technology (unless a technological innovation leads to a dramatic change in the price/performance ration; in the case of the electric car, this could be a revolution in electricity storage technology for example). Any support from the state should be limited to keeping open the options through basic research and some pilot projects in such cases, only if the technology shows significant positive "external effects".
- The cost curve of the new technology roughly fits the curve for the standard technology as soon as certain scale degression effects set in; at later stages of the life cycle, the cost curve can even fall below the standard technology curve (curve A2). This constellation is probably typical for many environment and energy technologies that are, presently, still stuck in the "innovation trap" (see above, chapter 5). For energy-efficient technologies, the capital costs are typically higher and the operating costs lower than for conventional technologies (e.g. heat pumps compared to gas heating systems). The 'economies of scale' are, therefore, of strategic importance for market introduction, especially concerning investment costs. State subsidies for market introduction ensure that

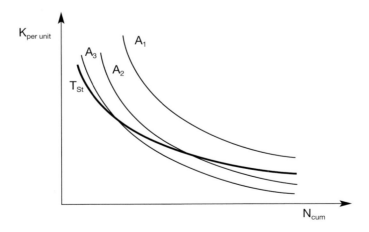

Fig. 7.1 Learning curves for energy-efficient technologies compared to the standard technology (T_{St}). ($K_{per unit}$: Unit costs, N_{cum}: accumulated production volume)

these barriers are overcome sooner and the new technology supersedes the standard technology. The subsidies must cease as soon as the threshold of economic viability is reached.

– The cost curve for the new technology runs to lower costs than the standard technology shortly after market introduction (A3). State action is not required in this case, since the enterprises themselves raise the initial market investments (gas turbine vs. coal-fired power plant). However, such technologies are often based on state-funded basic research (e.g. in materials science and technology).

7.3.1
The Dutch model

Any policy of promoting energy-efficient technologies must meet a number of *conditions*. First, it must be clear what technologies are promoted. An analysis of the technological options is not necessarily needed for compiling a list of the technologies supported (Ashford 2000). Yet, such an analysis for technologies that have still to be developed is found in Blok et al. (1995). In The Netherlands, the technologies already developed by companies appear in the so-called *Energy List* of the programs EIA and EINP[1] (Economische Zaken 1997, 2001a, 2001b). Wortmann (2000) produced a list of options concerning subsidized energy savings for households in Denmark and the UK and proposes similar measures for Germany.

When the list of technologies to be promoted is drawn up, economists usually ask how the state can know which technologies should be on this list. The procedure commonly used in The Netherlands is that the state publishes the date when the list is going to be prepared, at which point enterprises submit the technologies they propose. Then it is for the state institutions to prove that a technology does not belong on the list. This is a task for technologists and legal experts. A technology organization examines whether, from the ecological perspective, the technology proposed is better than the current technology. Arguments for this have to be given by the entrepreneur when he submits his proposal. Thus, the technology organization has only to inspect these arguments. Apart from this, legal experts have to subject the proposal to a formal examination, which will accept or reject the technology if it complies or violates, respectively, the legal requirements. In short, the procedure is designed in such a way that the state does not have to know what the list should look like. It only has to choose a procedure that gives an incentive to enterprises to volunteer their information in exchange for an advantage, i.e. the subsidies as a reward (for the lesser externalities of the environmentally sounder technologies as compared to the current technology). Furthermore, there must be an independent institution that carries out the examinations.

As the lists are revised annually, technologies that cease to be particularly good for the environment will fall by the way. Thus, the support is automatically limited to the period for which a technology is of particular benefit to the environment. If the learning curves for newer technologies are steeper than for old technologies, sup-

---

[1] *Energy Investment* **Allowance** *and Energy Investment* deduction regulation in the *Non-Profit* sector. The latter also includes agriculture.

port for them could be limited to the time when the cost differentials are still unacceptable in terms of regulatory law.

For subsidies that are paid in the form of tax allowances, there is no additional administrative effort required beyond the general administration of tax payments.

The aforementioned Dutch programs, EIA and EINP, provide the following funds (see table 1) for getting beyond a 'business-as-usual' scenario concerning the potentials for energy savings (see especially chapter 5).

Table 7.1 Funds available for subsidies towards energy saving investments in The Netherlands (In mil. €)

Year	EIA	EINP	Total
1997	61	1.14	62.14
1998	82	12.27	94.27
1999	105	16.36	121.36
2000	150	22	172
2001	186	25	211

EIA: Energy Investment Allowance *(EnergieInvesteringsAftrek)*
EINP: Energy Investment deduction in the Non-Profit sector

Source: www.ez.nl/subs; Staatscourant; Effectiviteit Energiesubsidies 2000

The burden on households due to the Dutch energy tax may be illustrated by the energy bill for an individual household. For gas, it is summarized in table 7.2; the corresponding numbers for electricity are listed in table 7.3. Heating and cooking appliances are fueled by gas in this household, which is typical for The Netherlands, but not for other countries. The household considered here is not quite representative as the household income is slightly above average, and there are three children. The example only serves to demonstrate the magnitude of the burden on households and the national budget due to this energy saving model.

The policy of raising an energy tax was first implemented in 1996. The second column of the tables shows the fluctuations in gas and electricity consumption. The taxed gas volume also fluctuates. The difference between columns 3 and 2 is regarded as unavoidable consumption. The tax per unit consumed according to column 4 was increased over time from a low initial level. For electricity, this increase only took place from 1999. These extra charges resulted in increased tax payments by the household considered, as shown in column 5. As this is a five-person household, these tax payments are divided by five to arrive at the numbers in column 6. The results are summarized in table 7.4.

The subsidies for the year 2000 are equivalent to about 0.1% of the government budget and amount to less than 0.05% of the GDP, without using up all available funds. The ceiling for these funds is adjusted if it turns out to be insufficient. Therefore, this also represents the magnitude of the *possible* tax increase that can be attributed to the energy policy. The following considerations help to explain why

Table 7.2 Gas consumption (m3) and energy taxation (EURO) of a household in The Netherlands

(1) Year	(2) Gas(m³)	(3) Taxed m³	(4) Tax/m³ in Cent	(5) Tax in EURO[a]	(6) Gas-energy tax/ person[b]
1996	3594	1860	1.47	27.39	5.48
1997	3420	2620	2.37	62.05	12.41
1998	2965	2179	2.03	83.88	16.78
1999	3135	2343	6.10	142.87	28.57
2000	3011	n.s.	n.s.	190.50	38.10

Table 7.3 Electricity consumption and energy taxation of a household in The Netherlands

Year	Electr. KWh	Taxed kWh	Tax/kWh in €-cent	Tax in EURO	Electricity tax /Person
1996	2464	1355	1.34	18.17	3.64
1997	2424	1624	1.34	21.78	4.35
1998	2421	1634	1.34	21.91	4.38
1999	2540	1748	2.07	36.20	7.24
2000	3069	n.s.	n.s.	76.96	15.39

n.s. not specified
a Column (4) multiplied by column (3)
b Column (5) divided by five persons

Table 7.4 Key figures for the Dutch model of an eco-tax

Energy tax bill per person in 2000:	€ 38.10 + 15.39 = € 53.49.
Energy tax bill per person in 2000, incl. VAT:	€ 62.85.
Energy tax revenues for 17 mil. residents in 2000:	€1068.48 mil.
Energy tax revenues as a percentage of the government budget:	0.6%
Expenditure for subsidies:	€ 172 mil.
Income tax reduction:	€ 1068–172 = € 896mil.
Gas consumption decrease[a] due to eco-tax, 1996-1999:	1.4%
Electricity consumption decrease due to eco-tax, 1996-1999:	4.4%
Cost efficiency of EIA subsidies[b], in kgCO$_2$/€:	9–19
Reductions due to EIA subsidies[b] in kton CO$_2$:	599–1206

a See http://www.ez.nl/Persberichten/persberichten2001/2001113.htm 9/27/01.
b for 13 selected technologies, excluding arbitrage effects; according to *Effectiviteit Energie-subsidies 2000*, p.60.

the amounts as a percentage of the government budget or the GDP are so low: The costs of energy consumption as a percentage of the GDP are low in any case. Concerning investments, on the other hand, any additional costs covered by subsidies must have arisen from an increased environmental soundness. According to an agreement reached in the WTO, such subsidies must not exceed 20% of the total amount of investments. Under the Dutch regulation, the actual number is about 15% with the additional costs mentioned above as an upper limit. Since the gross investments amount to ca. 20% of the GDP, subsidies can account for 3% (15% of 20%) at most. The actual figure of 0.05 % is much less, however, because the additional costs often exceed the 15%-mark, or because better technologies are frequently not available. In both cases, there will be no demand for subsidies.

One can avoid energy tax payments to the extent to which one saves energy while enjoying the income tax reduction at the same time. Against the reduction due to the eco-tax, rising incomes lead to increased total electricity consumption because of the greater number of electrical appliances purchased. Falling prices for the energy functions, light, heat and power amplify this effect. In The Netherlands, these functions became 20% cheaper over the period 1982 to 1997 (Bleijenberg and van Swigchem 2001). The Ministry of Economic Affairs in The Netherlands is currently implementing the following new programs: On 12 February 2001, the 'Energy Savings through Innovation' program was inaugurated. This program makes available € 16.36 mil. for projects in the phase between having an innovation idea and its market introduction, for investments and innovative applications of existing technologies.[2] In addition to this, one million € was allocated from industry sources, to be used for feasibility studies and knowledge transfer, because of the extraordinary interest in the program.[3] € 30.91 mil. were made available for investments that help to reduce greenhouse gas emissions and lead beyond current practices. Projects yielding more reduction per € of subsidies are treated preferentially.[4]

The Dutch regulations (see Economische Zaken 1997, 2001a, 2001b) stipulate that the subsidies towards the additional costs of such technologies that progress beyond the 'state of the art' and appear on the so-called energy lists must be paid out through tax deductions granted by the internal revenue service. In this way, four different controls are provided: The auditors have to confirm to the companies that everything is properly accounted for. The tax administration service inspects these accounts and, in some cases, the company itself. Competitors can bring a case to the EU and the WTO if they feel victimized by market distortion or protectionism. Naturally, 'good governance' is a precondition in all these areas. Notwithstanding all the problems in connection with certain subsidies, these fourfold controls are probably sufficient to guarantee proper procedures. Consequently, this model is hardly susceptible to abuse.

In addition to this, the ministries of finance and economy have carried out joint and, in the case of the latter, independent evaluations of the model (see Tweede Kamer 2001). They came to the conclusion that the arbitrage effects, which are defined as investments that would have happened anyway, even without the subsidies, amount to 50% of the total sum of money invested. The reason is that there are too many standard technologies on the energy lists, i.e. technologies that would be used anyway. As a result, the demands put on technologies on the energy list will be higher in the future. The standard technologies will thus disappear from the energy lists. As soon as

---

[2]   http://www.ez.nl/beleid/home_ond/energiebesp/firstlevel_index.html, 9/27/01.
[3]   http://www.senter.nl/energiebesparing/pb17092001.htm, 9/27/01.
[4]   http://www.ez.nl/Persberichten/persberichten2001/2001128.htm, 9/27/01.

the technologies eligible for the list have to be environmentally sounder than the standard technologies, arbitrage effects or buyers' returns can only occur to the extent that the buyers would also have bought at a lower subsidy. But if the subsidy is limited to the additional costs, any returns are excluded in principle. The present situation is that only 22% of the additional costs for a certain selection of technologies (see *Effectiviteit Energiesubsidies 2000*, Table on p. 59 are covered by the EIA subsidies. Hence, as soon as the standard technologies have disappeared from the list, the subsidies for the remaining technologies should be increased, in relative terms, to pay for a larger share of the additional costs. Then, arbitrage effects can only occur to the extent to which there remain difficulties in distinguishing between standard and environmentally sound technologies, or in determining precisely the additional costs. It can be difficult to recognize the technologies that are already standard technologies, because some investors choose environmentally sound technologies in the absence of subsidies, while others do not. The reason might be that the former regard the introduction of an eco-tax or certificate systems as more likely or more expensive than the latter. As soon as these expected costs are taken into account, some of the environmentally sound technologies become profitable even before an environmental policy is introduced; they *already* are standard technologies, while the environmentally less sound technologies *still* are standard technologies for entrepreneurs who do not expect the introduction of environmental policy. In short, heterogeneous expectations can make it difficult to identify the standard technologies.

The regulations existing on the EU level have been summarized by Blok et al. (1996). In Austria, 12% of the revenues from an energy tax on gas and electricity, introduced in 1996, were made available to the federal states, for implementing energy-saving and environmental protection measures (see EVA 2000). In other European countries, especially in Germany, such subsidies have hardly been used so far, although clean technologies produce positive or, at least, less negative externalities.

7.3.2
General considerations

Subsidies are unpopular both in the eyes of the general public and in the scientific discussion. The probable reason is that they are widely identified with protectionist subsidies rather than with sensible investments in research, infrastructure and the environment. However, the well-known fact that most subsidies are protectionist subsidies is the result of political processes. The instrument as such cannot be blamed for this, since similar things can be said against tax reliefs and economic regulations (which are protectionist in many cases). This is more connected with the fact that the losers in a structural change are always better organized than the potential winners. The first are affected directly, and right from the beginning, in their economic "acquis", whereas the winners may not even be known yet, and the profits are accrued only later. In a similar way, "producers" (including the trade unions) are always better organized than consumers. The former usually obtain their income in one sector; as consumers they spend in many sectors. Hence it is (economically) rational for the consumer to focus on improving the income situation.

In chapter 8 we discuss strategies on how one could shape policies more oriented to innovation. The present chapter concentrates on subsidies for energy saving technologies that differ from standard technologies in the sense that they will not be purchased without subsidies.

Subsidies for energy saving investments reduce the fixed costs of enterprises. They offer an incentive for trimming down the energy input coefficients in such a way that their effect is saving energy rather than saving labor. This leads to an increased factor productivity and hence to a compatibility of innovation and environmental policies (see section 6.2.6). The corresponding reduction of the marginal costs results in lower monopoly prices so that the conflict between competition and environmental policies is defused (see section 6.2.2). However, since lower marginal costs lead to higher industrial production volumes, the positive environmental effect is partly cancelled out.

In the negotiation process with trade unions, lower marginal costs and prices tend to lead to real wage increases, more job vacancies, more tensions in the labor market and a slightly lower rate of unemployment, whereby the conflict between environment and employment policies is defused (see section 6.2.1). But more employment also gives rise to more pollution and, again to a partial compensation of the effect of choosing an energy-saving technology. (These effects are explained in detail in Ziesemer (2000)). Still, the efficiency improvements of 4% p.a., which are needed for achieving the "2000-watt benchmark", are not expected to be compensated for or even overcompensated for in any of these cases.

The same applies to the risk that the "double dividend" might not materialize. Subsidies shift this risk from employment back to pollution, since higher employment numbers lead to increased production and the increase of emissions thus triggered counteracts the effect of a reduction in the energy input coefficients. Insofar as the energy saving investments flow into international transport, international trade becomes cleaner and the conflict between environmental policy and trade liberalization is brought close to a resolution. If this makes international transport more attractive, on the other hand, the energy-saving environmental advantages are partly cancelled out again. In this case also, overcompensation is unlikely under the conditions assumed here.

As enterprises receive support grants, and their choices are not restricted, there is no reason for them to move abroad because of this policy. Since this measure is based on adopting certain technologies domestically, it will not conflict with development policies; it does not increase transport costs and any technical progress will usually benefit the consumers in both countries. The enterprises will even gain competitive advantages as lower prices at home lead to substitution and income effects, which redirect the demand abroad to the domestic market. This creates an incentive for foreign governments to practice a similar environmental policy.

The additional demand for energy saving technologies also produces an additional incentive for research in this area. Right from the start, an inventor can count on subsidies he will receive if his technology is particularly energy saving, compared to other technologies. Where new technologies are not yet widely known, a subsidy system can help in publicizing them. In the Dutch subsidy program EIA, at least 3% of the subsidized investors only heard of the technology that would be adopted later when they learned that the subsidy was available (see *Effectiviteit Energiesubsidies 2000)*. Aside from their economic effects, subsidies also have psychological effects on managers, who are uncertain in their decisions on an investment. The existence of subsidies signals to them that other individuals and institutions regard certain technologies as sensible and assume that a subsidy system supports certain investments (see Interdisciplinary Analysis 1998).

One often hears the objection that subsidies are of little effect. Subsidies are price changes and have the effects of price changes. Hence, the objection is identical with the assertion that price effects are weak. This can actually be true if the price elasticities are low. But if a market economy is the best of all practiced or even conceivable economic systems, the price mechanism must appear to be the best available mechanism for steering an economy – despite all market imperfections. If that is the case, one should place higher demands on subsidies than on price changes in general. The so-called arbitrage effects in the context of subsidies are the equivalent of consumers' surplus in the case of prices. In other words, the criticism that subsidies are not effective enough is based on excessive expectations concerning the effects of price changes. The mistake is made when the critics form their expectations, unless there are problems with implementation and administrative handling.

It is for the World Trade Organization (WTO) to check whether subsidies represent protectionist measures, especially whether they serve to substitute non-tariff trade barriers for customs duties. However, subsidies for infrastructure, research and development are not subject to suspicions of protectionism (GATT 1992). As already mentioned, the WTO regards 20% of the industry costs incurred through the adaptation of equipment to new environmental legislation as non-actionable (WTO 2000).

On the European level, also, environmental aspects must be taken into account when granting subsidies, at least if they affect technical developments (also see section 6.3.3).

Tradable permits offer an alternative, or possible supplement, to subsidies and taxes. The seller of a certificate enjoys a cost reduction similar to the case of subsidies, if the returns from selling certificates exceed the costs of reducing emissions. The buyer of tradable certificates experiences a cost increase comparable to the one due to taxation. If the certificates issued do not cover the existing emissions, enterprises come in for higher net costs for reducing emissions or for buying additional certificates. Therefore, the conflicts discussed above are similar under tradable permits. This is a critical disadvantage of trading certificates domestically. In the case of pure "grandfathering" of certificates, there is the additional factor that new enterprises are not entitled to be issued with certificates even if they use the same modern technologies as the companies that obtained certificates (concerning this aspect, see Nielsen et al. 1995). This is a (strategic) distortion of competition, which may explain why some companies take a quite positive attitude to an active environmental policy: They expect their competitors to suffer more from the environmental policy than they do themselves (raising the rival's cost).

Nevertheless, fixing the number of the certificates traded is the only way of laying down an international limit on emissions that cannot be dodged by means of other tax measures (Eizenstat 1998). Employing certificates on an international scale and using subsidies at home are complementary measures. Thus the outlines of a potentially three-layered system emerge: (i) Governments and, perhaps, enterprises buy and sell certificates in an international market; (ii) domestically, voluntary self-commitments are entered into, which include (iii) subsidy rules like those explained above. The subsidies and possible government purchase of certificates in the global market can be financed through taxes on polluter households (polluter-pays principle, see section 7.3.3). Such taxes are raised e.g. in Germany and The Netherlands. The aforementioned national rules can also be applied if an international system does not materialize, for instance because the developing countries, and therefore the US, decline to join in. Even in the pure, simplifying theory, we face the problem that coalition models only have the following result: "Whenever the cooperation of sovereign states is urgently needed, it cannot be counted on" (Böhringer and Vogt 2001, p.4). If this is the case, the international agreements required for tax and certificate solutions will not come into existence.[5] However, this does not affect the proposal presented here: of achieving some progress by means of subsidies.

If a mixed certificate/tax solution is ever introduced, subsidies will be complementary instruments. Certificate costs and taxes are raised per unit of energy consumed and therefore increase the marginal costs, subsidies lower the fixed costs and energy-saving measures reduce the variable costs. If such a policy does not produce enough effects, it would be even conceivable to raise an additional tax on energy consumption or emissions.

It should be clear at this point that using these instruments is not without cost, but that it is accompanied by a more even distribution of the costs of the environmental policy over all households and thus over all beneficiaries of the policy. The users can be integrated within the framework of voluntary agreements. In the Dutch EIA system, for example, 65% of the subsidized investments go to enterprises that are party to a voluntary agreement (see *Effectviteit Energiesubsidies 2000*). These considerations are implicitly based on the idea that saving energy is the central point of sustainability in environmental policy, because renewable energies can replace fossil energies only to a limited extent. There is a crucial objection to this view (Sinclair 1992): If the OPEC countries expect a price fall or weakening of the growth path because of a weaker demand for energy, forestalling the price drop by increasing the supply might be the best strategy. This would boost the greenhouse effect. However, one may object here that the increased supply would be faced with an inelastic demand. Consequently, prices would drop drastically, at least in the short term. It is a question of time preference whether OPEC would put up with this short-term loss of revenues. Sinclair shows that, according to Hotelling's Rule[6], the extraction rate grows and the greenhouse effect increases in the absence of extraction costs. If one models the extraction costs by means of a production function, assuming a steady state with constant extraction and interest rates, prices will fall

---

[5]   The climate conferences of Bonn and Marrakesh did not change US policy on non-participation for the time being, because the developing countries do not want to cooperate. Sanctions and initial endowments of certifi-cates have not yet been settled. Only when this has happened, will ratification decide whether there will be an international treaty.

[6]   According to Hotelling's Rule, the growth rate of the resource prices must equal the prevailing interest rate, if the extraction costs are zero. If the interest rate exceeds this growth rate, total extraction will ensue so that the income from it can be invested in the capital markets; if the interest rate is lower, extraction will cease, because later extraction and sales will yield higher returns.

along with the rate of progress in extraction technologies. In this situation of exogenous technical progress, environmental policy will have no influence on the rates of price change. Future research has to decide whether Sinclair's result is still valid then.

It is in the interest of Arab states to be able to approach this problem by increasing prices, as was the case during the oil crises. This generates pressure on tax increases, even if these were very modest. The incentive is shifted back to saving energy, however in a way tantamount to a tax increase, as far as the goals discussed above are concerned, with the revenues disappearing abroad and thus unavailable for the support of technologies at home. Therefore, this wish of the oil companies might not be fulfilled – unless the interests of Arab and American oil companies coincide and find the (perhaps decisive) support of the American government. This distribution contest and Sinclair's problem would probably have to be dealt with simultaneously in a differential game.

Another possible objection against the proposal to use subsidies for defusing conflicts of goals could be that structural change and consumer abstention are regarded as essential for achieving the necessary energy savings: If the measures really turn out to be too ineffective, it might be necessary to bring about structural change and consumer abstention through environment taxes. Still, this requires a shift in the balance of powers in the conflicts discussed above, which could actually happen if the environmental problems develop in a way that leads to a heightened awareness of their relevance. In any case, it will always be possible to supplement the proposed subsidies for positive externalities with taxes or tradable certificates.

7.3.3
Ways of financing energy saving measures

An obvious solution would be to finance the proposed subsidies' policy for energy saving technologies during the phase when the costs are unacceptably higher than for the standard technologies (see the initial paragraphs of section 7.3) from *reductions in protectionist subsidies*. The problem is that, although everybody knows that protectionist subsidies are inefficient, such subsidies are still in place for reasons of distribution policy. Any demand to reduce them may be reasonable, but hardly realistic. The same is true for all other suggestions concerning the reduction of inefficiencies in the tax system. Therefore, our strategy does not rely on such approaches.

A more realistic way, even if unpopular among economists, is the *"lawnmower method"*. Its idea is to find out, for each political decision, what percentage of the relevant budget is needed for implementing that decision. Each government department is then obliged to save the same percentage of its budget. This method has the advantage that it forces the departments to inspect their budgets continuously, to implement savings and to establish priorities in this way. The drawback of the method is that it remains entirely unclear whether one is coming closer to optimum taxation or if one is moving further away from it. For this reason, the method is not necessarily recommendable, even if it is absolutely possible as a practical solution.

A *recommendable method* must meet certain criteria: (i) It should be efficient in itself; (ii) it should ensure that as many people as possible *will actually be* better off, not that they *could just be* better off. Only if enough parties will benefit, can the necessary majorities be found in the political bodies. A crucial problem with the introduction of an energy tax, and hence with the polluter-pays principle, is precisely that every citizen profits from an improved environment, but only the polluters pay for it: the enterprises as direct contributors, the workforce indirectly, through a weakened demand for their qualifications, and perhaps the politicians by losing votes. Compared to the status quo, the polluters will probably be worse off and therefore resist any energy tax or similar measures.

If, on the other hand, households, as co-originators and victims of environmental problems, are made to pay for environmental improvements, it can be ensured that the new burdens do not exceed the utility gains for a large number of people. Then the polluters have no reason to oppose an environment policy any more. The question remains as to how much tax one is prepared to pay for how much environmental policy. According to Böhringer's and Vogt's (2001) interpretation of poll results, the readiness to pay is not very strong. However, the poll results shown in table 1 of their paper leave ample room for interpretation, especially concerning the tax rates into which such results would translate. It is a particular strength of the principle of making the beneficiaries pay that they do not get more improvement in the environment than they are actually prepared to pay for. Free-rider behavior is thus excluded. An environmental tax on energy consumption or on the emissions related to it *for all polluting households* would be constrained efficient in the sense that it would internalize a negative externality. Furthermore, this internalization would affect everybody who profits from an improved environment. This is equivalent to a mechanism in modern microeconomics: A positive externality/ reduction of a negative externality (in this case a clean technology) is subsidized and a negative externality (emissions from households) is taxed in order to finance the subsidies. Hence, we recommend this method. As households are victims and polluters at the same time, it means that the polluter-pays principle applied to households is fully effective. Ziesemer (2000) provides the theoretical proof that this method reduces unemployment. The polluter-pays principle is not applied to enterprises, on the other hand, because of the market imperfections mentioned above.

The issues surrounding intergenerational distribution, which were touched upon in section 3.2.3, cannot be resolved conclusively in this context. If the greenhouse effect is stalled, all present and future generations, as well as the actual polluters, will bear all the costs. This situation would be distribution neutral. If nothing is done, future generations will carry the entire burden caused by themselves and the present generations. As the greenhouse effect can only be moderated, but not prevented, and because future generations will likely bear the still unpredictable, remaining burden of the climate problem while the present generations only pay the prevention costs, which are still low, we can see what the distribution will be: The present and future generations prevent less than they have caused and, therefore, pass on part of these externalities. Hence, any additional tax burden on households in connection with our proposal developed above can only achieve a reduction of the distribution gains they attain at the expense of future generations. Only if one chooses to use the option of "doing nothing" as a criterion, do present generations lose out through our proposal. This criterion can hardly be justified, however, and certainly not with the polluter-pays principle, from which we depart only where it appears politically unavoidable because of conflicts of goals.

This proposal, which is quite similar to the Dutch model, must entail statutory, binding payments that must comply with the legal requirements discussed above. If one is not prepared to finance these subsidies through cutbacks in preservation subsidies, increasing the state's share could be unavoidable in the end. Such an increase is conceptionally justified because the internalization of externalities through taxes and subsidies is exactly what economists recommend. However, the state's share must be kept unchanged by reducing the state's expenditure in other areas. Hence, in countries where the state's share is regarded as too high, economically unjustified expenditures and revenues must be tracked down instead of running a policy that is inefficient for both the environment and the common good. The exception is the establishment of markets for tradable certificates, where the state budget is not involved. The disadvantages of the certificate trade were already pointed out by showing that all the conflicts of goals discussed in this chapter actually exist on the side of the buyers, even if to a lesser degree in the case of 'grandfathering'.

7.4
Action field energy efficiency in industry: Self-commitments as an instrument for the rapid diffusion of the "best available technology"

7.4.1
General considerations

While subsidies are recommended specifically for accelerating the market introduction phase, self-commitments are more suitable for other phases, as the following considerations will show.

In some European countries, unilateral and joint self-commitments, including so-called agreed and negotiated solutions, have become established instruments of environmental policy. Self-commitments exist in most countries of the European Union. Most of the (ca. 100) were entered into in Germany and The Netherlands, but there exist 10–20 self-commitments in other countries like Austria, Denmark, France, Italy and Sweden, according to a review for the year 1999 (OECD 1999). Internationally, one refers to 'voluntary' or 'negotiated agreements'. Most self-commitments concern waste or climate issues. Their effectiveness is controversial: Industry cites as an advantage their flexibility, which makes possible lower implementation costs, their rapidity (no legislation) and their scope for dynamic development when conditions change or new objectives arise. Environmental associations, on the other hand, are concerned that their goals are less than ambitious and the achievement of these goals cannot be monitored sufficiently. Political regulators perceive risks for competition introduced by industry-wide cooperation with the state.

Politicians often rely on this instrument when statutory regulations are difficult (for instance if a substance or material would have to be outlawed in all countries of the EU). In addition to this, industry must also have an interest in entering such a commitment, e.g. in that it would avoid a (future) statutory regulation.

Independent of the specific motivations of the parties concerned, European studies have pointed out that one should distinguish between different types of such self-commitments (e.g. Rennings et al. 1997):

i) Self-commitments that merely codify the 'business as usual' economic scenario. This signals the factual absence of any environmental policy in the sense that such commitments only reaffirm plans that exist anyway.

ii) Cost-free self-commitments with environmental policy objectives, e.g. the replacement of CFC with another inexpensive substance.

iii) Self-commitments involving tax and subsidy instruments. Rennings et al. do not quote any example in their study on self-commitments in Germany. This is the most desirable form, in their opinion, when the low-cost options are exhausted – which is generally accepted to be the case today. Obviously, such self-commitments are complementary to our subsidies proposal. They exist in Denmark and The Netherlands, in connection with subsidies and energy tax exemptions (Blok et al. 2001b).

iv) Self-commitments of a contractual structure entailing obligations enforceable by sanctions (not considered explicitly by Rennings et al.). An interesting example is the Swiss Federal Law of October 1999 On the Reduction of CO_2 Emissions. According to this, bulk consumers and energy-intensive sectors that are threatened in their international competitiveness can be exempted from the CO_2 levy, if they commit themselves vis-à-vis the Federation to restrict CO_2 emissions. To this end, an action plan must be presented which covers the period up to 2010, and regular reports have to be produced. If the company or sector fails to meet these commitments, the CO_2 levy must be paid subsequently, with interest.

The last two examples show that even in the case of self-commitments, the context matters a lot, especially the complementary instruments. Beyond that, several European studies (Lautenbach et al. 1992, Rennings et al. 1996, ELNI 1998, Knebel et al. 1999, Frenz 2001, von Flotow and Schmidt 2001) allow the identification of the following preconditions for the success of a voluntary agreement:

– Clarity and specific objectives, including the definition and differentiation of the underlying problem, as well as quantities that can be measured and operationalized, to be realized by enterprises.

– Awareness of economic consequences for the enterprises. These can rarely be determined precisely (particularly since the costs of the alternative are not known). Consequently, the assessment by the parties concerned is of crucial importance: The higher the assumed costs, the lower will be the preparedness to take voluntary action, or the bigger the incentives have to be.

– Heterogeneity and number of the enterprises concerned, the value chain and the market structure: The more fragmented the interests and the larger the number of participants, the more difficult will be the negotiation and realization of a self-commitment.

– Transparency and monitoring, mostly by a neutral third party (with regard to external claimants' groups and to the parties concerned), for measuring and documenting the attainment of goals;

– Negotiation power, power of persuasion and implementation capacities of the associations concerned vis-à-vis their member companies, in whose name the agreement is entered.

These general considerations will now be concretized by means of an important agreement that is of particular significance for energy policy, in order to derive conclusions from it for the accelerated diffusion of best available technologies.

7.4.2
Self-commitments for the reduction of CO_2 emissions

Apart from the Swiss example mentioned above, there are a number of other cases of self-commitments that aim directly at the reduction of CO_2 emissions, among them e.g. the agreement between the European Automobile Manufacturers' Association (ACEA) and the EU Commission. According to that agreement, the CO_2 emissions of automobiles newly registered in the year 2008 must not exceed 140 g CO_2/km on average (meaning a reduction of 25 % compared to the 1995-level). An interim measurement is to be made in 2003. Also on the EU level, there are voluntary agreements between the European Commission and manufacturers of electrical appliances. For instance the manufacturers of TV sets and video recorders have undertaken to ensure that the stand-by losses of their products do not exceed 6 W. For most areas, however, e.g. for industrial energy consumption, agreements on the EU level are considered to complex.

Still, in several countries there exist agreements concerning the efficiency of industrial energy consumption. The earliest long-term covenant exists in The Netherlands, where a large majority of industrial enterprises has promised to reduce specific energy consumption by 20% over the period 1989 to 2000. On average, the enterprises have actually achieved this target, although this success must be attributed mainly to the large share of the chemical industry in total industrial energy consumption. Further studies have shown that about a third of the reduction in specific energy consumption is an effect of this self-commitment. The Dutch examples are distinguished by intensive state efforts (communication, subsidies, etc.). A different kind of self-commitment is found in Denmark, where the CO_2 tax existing there also applies to enterprises. As a company enters into this self-commitment, it becomes exempt from this tax. This system is similar to the one in Switzerland.

In Germany there is, first of all, the self-commitment of the Confederation of German Industry (*Bundesverband der Deutschen Industrie,* BDI), agreed in 1995 and renewed in 2000, which now prescribes a CO_2 reduction of 28% by 2005 and a reduction of all six so-called Kyoto gases (CO_2, CH_4, N_2O, SF_6, HCFC and CFC) by 35% in total by the year 2012, all compared to the reference year 1990. To avoid a quota system, the agreement was supplemented, in 2002, by additional measures to promote combined heat and power plants. The EU commission is in the process of examining if these undertakings justify the exceptional eco-tax rules applying to industry. The binding character of the commitment is a crucial criterion in judging this matter.

For the chemical industry, the results from the first phase (till 2000) have recently been studied in detail (von Flotow and Schmidt 2001). Most interestingly, a target set for 2005 – reducing CO_2 emissions by 30 % compared to 1990 – was already met in the year 2000. The results were evaluated by the *Rheinisch-Westfälisches Institut für Wirtschaftsforschung* (RWI). The main industry measure was a more rapid introduction of known modern energy technologies (the integration into a complex, joint chemical production process, can also be regarded as an innovation here). Typical examples were the use of gas and steam turbine plants (instead of coal-fired ones), the construction of residual-water turbines and the installation of combined heat and power systems instead of regular boilers. This was made possible, depending on the individual enterprise, through longer pay-back periods (which promote more capital intensive investments), preferential inclusion in investment budgets and a generally stronger awareness of climate policy effects – encouraged by CO_2 reporting obligations (e.g. in the environment report) – when making investment decisions. Apart from the "hard" technology decisions, the self-commitment process also had a learning and information effect that, even if difficult to prove in any individual case, led to a change in management practices. This was made easier by the fact that the relevant circle of corporations affected was relatively small, and that the Energy Committee of the Confederation of the Chemical Industry (VCI) provided a ready platform for the exchange of experiences and information, and for benchmarking.

If self-commitments are structured appropriately, they can speed up the application of best available technologies for achieving objectives agreed upon, not least because of their awareness and learning effects. This is a better result than a business-as-usual scenario, for two reasons: First, the environment agencies in many European countries can only demand legally (and often under very limited conditions) that enterprises must retrofit using the respective state-of-the-art technology. In many cases, however, the best available technology is far superior to the latter, in terms of its environmental performance. Second, the renewal brought about by the diffusion processes thus initiated happens more rapidly than the normal replacement cycle, which is dominated by the consmption rate of fixed capital.

For the considerations presented here, this means that voluntary self-commitments under innovation aspects do accelerate the diffusion of sustainable energy technologies, especially if the new efficiency standards are to be tested by early adaptors (see section 7.3) before becoming the mainstream in the sector concerned after a short period.

7.5
Technology Procurement

While the buying decisions of the (final) consumer are based on limited information, and are mostly intuitive, organizations have a more or less clearly defined process for such decisions, which are dominated by value-for-money considerations. Private organizations need specific incentives (positive or negative) to enable them to depart from the given market ratios, whereas the buying decisions of state organizations are also guided by other goals, as already mentioned, as long as this does not lead to market distortions. Hence, a public procurement policy, as well as a private one, could also create a demand-pull for sustainable energy innovations. As soon as the market standards have been established in the public sector, they also apply to the private sector (e.g. in the construction industry). So far, however, bureaucratic budget prescriptions and general risk averseness hold back the full use of this potential. Even large corporations are affected by this. There have been numerous efforts to use public procurement policy for promoting innovation and environmental protection, not least in a number of EU programs. Still, there seems to be little practical progress.

Some countries have even established "efficiency funds" for inducing innovative procurements in the energy and environmental sectors. The central problem is that more energy-efficient products are often more expensive to buy, but cheaper to run, in terms of operating costs. The bottom line is that they are actually cheaper, in the end, but buying them would still violate traditional procurement principles. It is obviously difficult to manage such processes without reforming the public budgets from public financing to business accounting methods. For the time being, we have to assume that specific measures are needed to overcome this barrier.

One possibility in this respect is the cooperation between different parties concerned, to achieve more transparency, more precise requirements and more purchasing safety. Of late, such collaborations, initiated by environmental organizations, are organized by energy agencies. In support of the development and market introduction of an energy-efficient product the concept of technology procurement is applied. Market-relevant buyers are brought together in a group and develop, with the help of experts, the requirements of the product, for instance with regard to its energy efficiency, its price and its noise emission. The prototypes developed by the manufacturers are assessed in a competitive process. The manufacturer of the winning product is rewarded with a minimum purchase guarantee given by the buyers' group, and by being publicly commended by the agencies.

As a pilot activity, which is supported by the EU, European energy agencies presently run the "energy+" program for fridge/freezer combinations. More than 100 organizations, including environmental and consumer associations, representing 15,000 retail outlets and over a million European households are participating (Ritter and Amann 2001).

Demands were made on manufacturers' products, especially in relation to their energy efficiency. The maximum energy consumption permitted was 42% of the European average consumption of comparable appliances (i.e. about ? of the figure for the "A" class). Within two years, 16 fridge/freezer combinations from four manufacturers fulfilled this requirement (see www.energy-plus.org). The best candidates are even more efficient, with only 33% and 35% of the power consumption, far better than the highest European efficiency class.

Such an approach appears suitable for other energy innovations, too: solar power plants for electricity production (see section 7.8) or microturbines and fuel cells for the combined heat and power supply of public facilities (including hospitals, army barracks etc.) could be examples for possible applications of such project collaborations for technology procurement.

As we have seen, technology procurement could play a larger role for the market introduction of sustainable energy innovations, for markets do not emerge spontaneously, but need to be developed.

7.6
Action field energy efficiency in households

7.6.1
Sustainable energy supply and the sovereignty of the consumer

There is another conflict of goals that should be pointed out: The idea that "the customer is king" and steers the market economy through his demands is much too simplistic (as is the opposite model, which paints the consumer as the weak-willed object of "secret seducers"). Through product and service innovation, opinion forming and behaviour-influencing marketing, exploring social trends etc., suppliers can create desires and influence decisions. However, they compete in this with each other, with the result that their efforts are partially neutralized because, as everybody knows, the consumer can spend the same money only once – and he learns from experience and a broad supply of information, communicates these to other consumers and is free to accept or refuse an offer. Thus, nine out ten new products or services are doomed to failure.

But the findings are unquestionable (see e.g. Leittschuh-Fecht 2001): On balance, the environmental effects of private consumption, (the "ecological backpack") are greater today than they have ever been. Efficiency gains in energy consumption as well as emission reductions were overcompensated by growth or increased consumer comforts. This growth varies for different sectors, but the overall picture remains largely the same. Another factor is that energy use in households is particularly inefficient (Knoop and Steger 1998). Reasons for this are the immanent "inertness" of buildings and their fittings (see section 4.2), which does not allow the immediate application of the most efficient technology, lack of information in some cases and, to a larger extent, the fact that the consumer disregards environmental aspects in the concrete buying decision and instead decides for prestige, lifestyle, comfort, or just for the best price (despite the environmental awareness often articulated in opinion polls).

Several explanations are given, most of which amount to the assertion that the consumer does not know the consequences of his decisions and, even if he knows them, believes his individual decision can only effect marginal improvements, as long as the same decision is not taken by everybody else. As he cannot assume that other consumers behave in an environmentally conscientious manner, he is not prepared to accept any disadvantages involved in such behavior (the free-rider dilemma). Hence, environmental protection is a relevant decision criterion for him only if it takes an "individualized" form (e.g. paying for residue-free foodstuffs, but not for the collective goods counteracting climate change or supporting the security of procurement: These goods, once they have been produced, can be used by everyone, whether he/she has contributed to their production or not).

Polluting behavior on the part of the final consumer is rather supported by the fact that this area, in comparison with the production sector, is hardly regulated. A regulation relative to the IPPC (Directive 96/61/EC, "Integrated Pollution Prevention and Control") could accelerate significantly the application of the best available technology, especially with regard to energy consumption, and force "energy wasters" out of the market (Knoop and Steger 1998). The problem here is, first of all, that a household (as the place of decision making and of consumption) cannot be regulated like a factory. The more fundamental problem is, however, that politicians are more prepared to enter a conflict with individual industries – especially if there is public pressure – than with the consumers, who, after all, are identical with the electorate. Industry has little interest in making an issue of pollution and resource consumption in connection with its products and services, not even in the sense of a mere "cognitive dissonance", apart from the exceptions where environmental soundness – however defined – is relevant to the consumer as a positive, distinguishing feature. Not even the environmental associations address their demands to the consumer, in most cases. Thus, the consumer appears to be a major obstacle against a sustainable (i.e. low) rate of energy consumption. And there seems to be no practicable or acceptable mechanism to bring him closer to meeting his responsibility.

There is no obvious way out of this dilemma, but there are certainly some pragmatic approaches for improving the situation. In principle, the instruments discussed below represent efforts to provide the consumer with better information, a better basis for his decisions and an enhanced ability to make his choices. A change of practices will lead to a change in the ideals that guide the consumer, and are reflected in advertisements, lifestyle magazines etc. In the same way as film stars have stopped smoking on screen, no one will wish to be seen as an "energy waster".

7.6.2
Greenpricing of eco-electricity

Only with the liberalization of the electricity market, was the consumer enabled to choose between different suppliers, who now differentiate the homogenous good "electricity". A niche market segment for high-quality, highly priced "green" electricity has developed. Some of the new competitors active there specialize in marketing electricity produced in an environmentally compatible way. In spring 1999, there were 63 such suppliers; in 2000, this figure had already risen to 153 (Wietschel et al. 2001).

However, despite this multitude of suppliers, there has not been a corresponding increase in sales, although the majority of the consumers shows a positive attitude towards green products and signals the preparedness to pay a premium price for electricity produced in an environment-friendly way. Various polls in the US, The Netherlands and Germany show that between 50% and 70% of customers would accept such offers (Langniß 2000, Bloemers et al. 2001).

Notwithstanding the dynamic development, the high acceptance and the preparedness to pay on the part of consumers, the market share of green electricity in Europe is still below 1% in most parts of Europe. Only part of the hard core of environmentally active consumers has so far been reached, but not the mass market. This is plausible because the large majority of consumers do not credit such electricity with any additional utility: Everyone enjoys the advantages of an improved environment, but only the buyers of "green electricity" pay for it.

Four factors – credibility, transparency, and an aggressive communication and pricing policy – are regarded as the determinants of the successful marketing of "green" electricity. Most consumers still do not trust the new products of established energy suppliers. First of all, the consumers must be convinced of the fact that the products are actually produced in an environmentally compatible way, and that they offer ecological utility. To overcome this obstacle, McKinsey recommends (Bloemers et al. 2001) confidence-building measures in the shape of alliances with institutions that are independent and recognized by the consumers, for instance environmental organizations confirming that clean electricity really comes from new capacities (instead of all other electricity becoming "dirtier", as often suspected).

Ecological credibility can also be conveyed by the voluntary issuing of certificates and labeling of the product. In this way, consumers are convinced that they are actually buying an ecological advantage for the higher price. The labeling has the additional effect of an improved market transparency, since it is more and more difficult for the consumer to find his way through the growing variety of "green pricing" products. A quality seal, awarded by independent institutions, must be considered a necessary building block for market penetration, although the experience with labeling has been mixed so far (see section 7.6.3).

Most of the green products offered by some established electricity suppliers can be regarded as part of largely defensive market securing strategy that is as such run at low levels of effort and investment (Wüstenhagen 2000). Eco-electricity would have a better chance if a dedicated, independent marketing concept were developed for it. For customers who are strongly guided by environmental principles, the reference to the ecological advantage may suffice; to reach wider sections of the population, the individual nature of the social utility has to be conveyed. Communicating mere factual information is not enough if one wants to explore potentials beyond the market niche; it will be necessary to address the customer in a more emotional manner. The aim should be to build a strong, independent brand, which is only possible through a major marketing effort.

The decisive breakthrough for eco-electricity will not be achieved through lower price differentials, but by combining it with additional services that guarantee a hassle-free management of the energy demands of the entire household, via the Internet. Elements of this are automatic metering and billing, advice on options for use and improvement, a 24/7 emergency service and more comprehensive service packages. This "convenience approach" in connection with new technologies appears more promising than attempts to narrow the price gap or just to rely on the ecology argument.

On the a whole, such offerings of specific eco-electricity can further the restructuring of the supply side towards regenerative energy sources, but it cannot be the only driving force behind such a development. For this to be possible, the obstacles would have to be lower, and the crucial preparedness to pay needs to be more advanced.

7.6.3
"Discriminating" labeling

Buying decisions for energy-intensive appliances are dominated by cost and brand considerations. Frequently, the main obstacle to choosing energy saving products is the absence of information of the kind that is needed for such a choice, or the considerable effort required to obtain such information. In this respect, both voluntary and obligatory labels make it possible for the consumer to reach an informed decision concerning energy consumption easily. They thus offer incentives for energy efficiency innovations and for speeding up the market introduction.

The International Energy Agency assumes that labels, minimum standards and voluntary agreements, if they are carefully planned and designed "correctly" (step-by-step instructions are given in OECD/IEA 2000a), give incentives for innovation and market transformation. To increase the sales of energy-efficient appliances and equipment for home and office, such instruments have been implemented in more than 30 countries worldwide. On the EU level, the following are in place at present:

- Labels and minimum efficiency standards for refrigerators and freezers
- Labels specifically for washing machines, laundry driers, dishwashers and lamps
- Minimum efficiency standards for electrical boilers
- Voluntary agreements for washing machines, TV sets, video recorders and audio equipment

Household appliances

The first Europe-wide labeling regulation was the one for household refrigerators and freezers, introduced in 1994. The appliances were divided into energy efficiency classes according to their power consumption (in descending order, beginning with "A" for the best models). The success of this measure in Denmark is described by Jänicke et al. (1998). The market share of A, B and C class refrigerators in that country grew from about 40% to ca. 90% between 1994 and 1997. The consumption labeling was supplemented by an additional instrument mix, e.g. special training of the sales staff, an energy tax combined with a CO_2 levy, energy saving campaigns and a payment for disposing of old appliances.

Above-average market shares of energy-efficient large household appliances have also been observed in Germany, The Netherlands, Austria and Belgium in recent years. Since the introduction of the label, manufacturers experienced strong pressure towards higher energy classes from retailers (see AEG 2001 and Waide 1999).

The European Commission (KOM (2000) 247 fin.) criticized the less than optimal implementation of the labeling guideline. In 1998, the energy label was still little used, but where it was found, it had a considerable effect. The Commission attributed the Europe-wide reduction in the power consumption of refrigerators and freezers by 27%, compared to 1992, to the statutory labeling obligation as well as to the minimum requirements concerning their energy efficiency (KOM (2000) 769 fin.).

An important criticism concerns the failure to update the label continuously to reflect technical progress; it remained based on the status of 1994 (Schlomann et al. 2001). Within the best class, there emerged an ever-widening range of performance levels that are no longer transparent for the customer. Thus, incentives for further efficiency improvements are lost and the best class just grows automatically. A "top-runner" approach has to ensure that the class system is regularly adjusted to the latest technical developments.

Brown goods

By now, efficiency labeling has been extended to the entire "white goods" sector, which includes washing machines, laundry dryers, dishwashers etc. "Brown goods", i.e. office, communication and home electronics, are still not included.

Since ever more households are equipped with more and more of such goods, their share of electricity consumption has been rising continuously through recent years. This calls for action, especially with regard to the in some cases considerable stand-by losses. Presently, the EU is trying to reach agreements with the manufacturers. (For an overview, see KOM (2000) 247 fin.).

Cars

The labeling of cars became reality Europe-wide in 1999 and was to be implemented in national law in 2001 (Guideline 1999/94/EG of the European Parliament and Council). Information about the fuel consumption and CO_2 emissions must be given in sales and marketing brochures. To give the customer the opportunity to compare different models, lists of the consumption figures of all new cars on the market must be published. Prior to implementation the Austrian Energy Agency (EVA) was instructed to study the possible structure and savings potential of the system (Fickl and Raimund 1999). A simple label showing the relative fuel consumption of the new car compared to the average for its class turned out to be particularly effective. According to EVA estimates, this instrument could help to reduce the fuel consumption of the entire car fleet by more than 4%. Because of the rising demand for transport, however, the expected efficiency gain will be overcompensated by volume effects. Therefore, technical innovations alone might not be sufficient for a significant reduction of the greenhouse gas emissions from the transport sector. Social and organizational innovations have also to be considered (see section 7.7).

Buildings

Presently, residential and service buildings account for more than a third of the total European energy demand. The biggest potential for savings lies in efficiency improvements in the heating of buildings. The implementation of technological innovations is a lengthy process in this field because the replacement rate for buildings is about 50 to 100 years. The different interests of investors and users are another obstacle in the way of reducing energy consumption. As it is usually the tenant who pays the energy costs, the owner is hardly motivated to invest in energy saving measures. So far, neither potential tenant nor prospective buyers can get standardized information on the energy consumption of a property and, hence, cannot take it into account for their renting or buying decision. Making available characteristic indicators in this respect can bring transparency and give incentives for reducing energy consumption. The Europe-wide introduction or harmonization of such an instrument was recommended by the Commission (KOM (2001) 226 fin.). In Germany, the Decree on Energy Saving demands an "energy consumption passport" *(Energiebedarfsausweis)* for new buildings. For the existing stock, such a pass is obligatory only in the case of significant changes to the building, whereas the European Commission recommends giving this information for all new buildings as well as for buildings brought to the market for sale or letting.

Because of the considerable inertness of the building stock with regard to renewal (see chap. 4, table 4.4), neglecting this area would have particularly grave consequences. Hence, a stronger energy-efficiency orientation of the regulatory framework for the construction sector (Energy Saving Decree, VOB etc.) can be extremely important for the long-term achievement of the goals we are dealing with. The body of regulations should, therefore, pass by such interests that prefer an approach more oriented to the status quo.

The experience with labeling is mixed, on the whole, especially because industry does not consider it in its interest to have a truly "discriminatory" labeling system, which would clearly distinguish "sustainable" and "non-sustainable" products. But only such a system would have a positive effect on (energy) innovations. On the other hand, it has to be said that it is not at all obvious what the differentiation criteria should be. This was already the case with the "simple" environmental labels, and it gets increasingly complicated with every additional dimension of sustainability. Furthermore, the multitude of different state-regulated and private labels (with and without the blessing of the environmental organizations) confused the consumer rather than giving him guidance.

To be of any interest in the context of sustainable energy innovations, a labeling system must fulfill two conditions:

- If a classification according to energy consumption is introduced, the standards must be dynamic in order to prevent increasingly large segments of a market from becoming "top performers" virtually by default.
- The labeling system must be massively promoted in the entire market, so that it becomes part of the consideration in every buying decision. This is probably the reason why the EU Commission often aims to establish these standards in the form of self-commitments of manufacturers (which is, quite possibly, in their protectionist interest, since, due to the higher energy prices in Europe, the products manufactured here are often more energy-efficient than imports from the US or low-wage countries). At the same time, ways must be explored as to how to prevent misleading labels.

7.6.4
"Public Private Partnership" and unconventional marketing campaigns

The very limited set of instruments gave cause to think anew about ways to influence households towards sustainable patterns of energy use, especially as the classic information campaigns about saving energy were regarded as not particularly successful.

One approach, most notably in Anglo-Saxon countries, is referred to under the keyword "Public Private Partnership". In such a scheme, public institutions cooperate with private enterprises in order to achieve a common goal. A relevant example in the context of the present study is the energy saving funds developed in the UK. By setting priorities, these funds offer possibilities for the targeted exploration of energy-saving potentials in the private and public sectors. They are financed through government grants and an extra charge on electricity bills. The energy industry is obliged to realize a fixed volume of energy savings in the residential sector. The actual type of the individual measures is at the discretion of the energy suppliers. The main measures so far in England and Wales were financially supporting energy-saving light bulbs, efficient household appliances and building insulation. In the years 1998 to 2000, 2713 GWh of energy were saved (Wortmann 2000). Similar programs exist in Denmark, were the energy saving fund is financed through an electricity tax. The target is to reduce the consumption by 75-80 GWh, annually, by measures encouraging conversions from electrical heating to gas and district heating systems, and for increasing the share of appliances of the EU-class "A" in private households and in the public sector.

In principle, such concepts are based on the idea that the "energy suppliers" should be less interested in selling ever more energy. The new business model of the "energy services provider" is supposed to compensate the loss of volume through additional value creation from energy-efficient services (and be more profitable).

Some companies with special interests in selling energy-innovative products are also trying to take this idea to the customer via the Internet, using the intensive ways of communication possible through this medium (Shreeve and von Flotow 2001). The Electrolux website, for instance, offers easy-to-use calculation models to demonstrate the savings that can be achieved by switching to energy-efficient household appliances, and it presents the "energy solutions" offered by Electrolux. This is absolutely in the commercial interest of the Swedish manufacturer, because the, in some cases, extremely economical, innovative appliances give an incentive to part earlier with old products that are, technically, still in working condition. In addition to this, the new, high-quality products also offer a better performance in other respects.

However, not only pioneering enterprises are trying to reach out to the customer. Environmental organizations are also beginning to set new standards with innovative campaigns. Supported by the WWF, the local environmentalists of the "BioRegional Development Group" of Beddington, London, managed to realize a large "zero-energy" development area, where the buildings and the entire residential environment (including energy-efficient bio-waste recycling) are designed to be sustainable. The local administration, a large housing association and one of the leading British architects, were persuaded to take part in this project. In this way, the group was able to prevail against competitors who were prepared to pay a higher price for the land concerned. Such examples (more detailed in Kong et al. 2002) even touch areas such as food preparation (gentle and energy-saving cooking methods).

The approach chosen by the *Öko-Institut*, the "Top10 Innovations" program for sustainable consumption, which is designed to run for several years, is somewhat confrontational, in comparison. Based on its "eco-balance sheets", the institute wants to position high quality, but affordable, "naturally ecological" products in the mass market. Their main focus is on energy-intensive mobility and household equipment. Large demand and order campaigns in cooperation with consumer organizations are intended to influence public and private purchasing agencies and create massive media attention. Following its launch in Germany, the program is to be extended to other EU countries.

However we assess the individual measures, the interesting point is that they depart from the conventional routes (and frontlines) and attempt to support technical innovations with social innovations and experiments. The learning effect induced in this way is probably more valuable than the immediate ecological effect of the effort. For, to repeat one of the central messages of this study, sustainability is a process of trial and error – especially in an area as "inert" and immutably structured as the energy sector – in which society and its members learn to establish sustainable processes.

7.7
Action field transport: Only "packages" can produce innovations

As explained in chapter 4, the fastest growth in CO_2 emissions and energy consumption is occurring in the transport sector. Furthermore, this sector is almost completely dependent on mineral oil. Hopes for a future reduction are limited because the dominance of the two very energy-intensive modes of transport, cars and motortrucks, increases with the degree of "detailization" in logistics and mobility. There may be innovative model projects, mainly in small and medium-size towns (vividly described in Leitschuh-Fecht 2002), to roll back the dominance of car and truck, but these are still "niches in a niche". Isolated technical measures concerning less energy-intensive carriers, such as railroads, are of little help in reducing the many-sided competitive advantage of car and motortruck.

Whole "packages" of mutually supportive measures are needed to reduce the energy-intensity of the transport sector through a combination of technical, organizational and social innovations. In the following, we briefly outline some examples.

The fuel cell package

The perhaps most important technical innovation is the prospect of a gradual replacement of the more than 100 year-old combustion engine by the fuel cell (which is even older, concerning the date of its invention). A number of technological breakthroughs, most notably in catalytic converter and membrane technology, have already made it competitive as a stationary engine; the same will soon be achieved with mobile versions. Even if some earlier euphoria has vanished by now, commercial buses will be available by 2005 and mass-produced cars with fuel cells by 2010. However, under sustainable-energy criteria this can be considered as progress only if the required electrolytic hydrogen can be produced using regenerative energy sources. Otherwise, the total energy consumption could even rise, depending on the option chosen for fuel production (more details in von Flotow and Steger 2000).

There are no links yet between the use of fuel cells and regenerative energy sources, but these can be created in various ways, e.g. through using biogas (see section 7.8.2) or additional wind energy capacities. (As hydrogen can be stored, this would solve the problem of interruptions in electricity production). In terms of industrial policy, this would amount to supporting the position of the fuel industry. The fuel suppliers argue for the direct use of hydrogen in order to avoid paying twice for retrofitting the gas stations. (The additional hydrogen retrofit of one gas station costs ca. 400,000 Euros). The automobile industry, on the other hand, favors the production of hydrogen from natural gas or petrol.

The urban transport package

Classic public transport systems are less and less able to satisfy our varied, time-dependent mobility requirements. Hence, we need an "urban mobility package" that allows a flexible choice among existing transport options, based on Internet information (more details in *Mobilität 2001*). This includes not only the Internet modality between the various (public) carriers (e.g. railroads, tramways, buses etc.), but the selective use of cars, edge of town parking facilities, car sharing and bicycle rental ("call-a-bike"). Public carriers will have to switch to networked minibuses running on synchronized schedules ("individualization of collective transport, collectivization of individual transport"). The important points are the easy availability of the information required to compose one's personal "transport menu" and the ability to use "smart" payment methods (creating the pressure to actually leave the car at home). The innovations leading to a sustainable energy consumption (compared to the car) do not, primarily, belong to the area of drive technologies. They are, rather, the "individualized" links between transport carriers and services made possible by ICT developments, especially by mobile access to the Internet. A project of the Swiss national research program *Verkehr und Umwelt* ("transport and the environment") (NFP 41) shows that a consistent urban transport package can make citizens give up their own cars without regarding this as a loss of comfort: While about 25% of all households in Switzerland do not own a car, this share is 40% in large cities. More than 80% see giving up their cars as a change for the better.

The rail intermobility package

All over Europe, fast long-distance trains have become accepted as an alternative to traveling by car or plane (up to a distance of ca. 300 miles). However, their use for business and holiday journeys is restricted by the focus on main traffic hubs. As was the case with urban transport, intermodular links are required here, enriched with services (from luggage service to "infotainment" on the train), and with transparent ways of access to information. The hesitant efforts at a more intensive segmentation of trains according to destinations and creating more linkage points for connecting options at railway stations should be extended on a massive scale. Most projects are technologically feasible already; Internet access on the train is only a matter of costs today. However, the service packages are still insufficient. The bottleneck lies in the organization.

The logistics intermodality package

A similar problem is encountered in the area of freight logistics. As complete trains transporting mass-produced goods become increasingly rare, and the punctual delivery of small parcels increasingly important, railroads too, depend on intermodality. The logistics concepts developed for the motortruck set the standard here, although they are more resource-intensive. However, handling times and costs are still much too high in rail transport. Necessary preconditions such as automatic couplings, individually motorized railroad cars and unmanned direction of freight trains or cars are, astonishingly, still technical demonstration projects instead of the standard. The same is true with regard to the advance information and accompanying information for shipments. Massive investments are required in a more decentralized structure of freight logistics and the complementary information and control technologies. This applies, in a similar way, to inland and ocean navigation.

For European transport and technology policy, this means a need to reduce the focus on supporting individual components and to test and improve the interaction between the components in large-scale model experiments with "packages". The procedures of technology procurement (see section 7.5) could be applied to the technical side of this approach. But the real challenge lies in the design and organization of the packages as pilot trials and demonstration projects. This is not just a question of support funds, but also of the choice of platforms for learning and innovation processes of highly diverse actors, who must be made to cooperate in spite of partly diverging interests.

The transport sector, in particular, demonstrates the necessity of a close link between organizational innovations and various technological ones in order to ensure the mobility of goods and people at a standard that is equal to the car and motortruck system, but less energy-intensive. This conclusion is based on the assumption that regional economic cycles and traffic prevention, even if they remain important objectives of transport policy, can slightly attenuate, but by no means reverse the trends towards (global) work-sharing and individual mobility.

7.8
Regenerative energy sources

7.8.1
General considerations

Regenerative energy sources comprise a wide spectrum from traditional, commercial technologies, for instance large hydroelectric power plants, to novel approaches such as the second-generation thin layer cells in photovoltaic technology, which are still at the stage of basic research. For other such sources, e.g. wind energy, there has been a rapid technological development so that that they will reach the threshold of economic viability in the near future; hence, the considerations in support of these sources, which we discussed in section 7.2, can be applied here. At this point we focus on the broad spectrum of regenerative energy sources that are at the stage of pilot projects or first demonstration projects (as described in section 5.3), including areas such as biomass, decentralized hydroelectric and, above all, photovoltaic energy.

This is a field of "learning by experimentation". Aside from the technology development itself, the reasonable combination of technologies and the integration into existing grids and infrastructures must be tested. One example of combined technologies is the biogas-driven (micro)turbine, which is particularly relevant for developing countries. The extent to which a grid can rely on "interruptible" regenerative sources (e.g. solar and wind energy) is an example of integration. In Europe, the limit was considered to be about 10%, but model calculations for California already predict shares of up to 30% (Williams 1994), if the grid is managed "intelligently".

Although the European target of reducing the total consumption by ca. 12%, and to produce 22% of all electricity from regenerative energy sources, by 2010 is ambitious, it is not supported, at present, by appropriate measures and programs on the European level. Even less foreseeable is a trend towards 50% by the year 2050, which would be needed for meeting the "2000-watt benchmark". There is still a plethora of individual measures and counterproductive trends. An example of the latter is the increased pressure on the electricity industry to use only the cheapest suppliers and fuel sources, as an effect of the (necessary) deregulation of the electricity market (see section 4.2.3; more details in Gather and Steger 2001).

Still, the national programs, if properly structured, can become Europe-wide "multiplicators" as soon as precisely those instruments that proved useful with regard to specific energy technologies in certain phases of the innovation cycle are applied at the European level. The aim is then, to reach, as quickly as possible, a degree of maturity that makes possible the wider market introduction with its economies of scale and the emergence of complementary service and qualification structures, so that we can at least begin to overcome the obstacles for innovation analyzed in section 5.5.

7.8.2
Technology-specific support measures

Based on the analysis presented in chapter 5, section 6.4 and in appendix 3, we recommend the following support measures in the EU:

Biogas for electricity production
Considering especially (micro)turbine development, biomass can be gasified in a decentralized system, too. The electricity can be fed into the grid or used for producing hydrogen. This decentralized gasification and "refinement" significantly eases the concerns otherwise connected with the use of (agricultural) biomass: the negative ecological effect of land utilization. It can be ensured that only land is used that is withdrawn from food production, anyway, in the context of EU agricultural policy. Since this will be a matter of scattered areas, no new monocultures will be created either.

The same approach can also be applied to refuse, sewage gas, landfill gas and mine gas. As this technology is still at the stage of demonstration plants and first series production today, we recommend limited support, for instance in the form of feed-in payments as in the German model. These payments can be reduced gradually as the technology becomes cheaper. Exploiting this potential is particularly relevant in connection with the technology sharing that is promoted through development policy measures (see chap. 9).

Decentralized hydroelectric power
While large hydroelectric power plants (from about 10 MW) represent an established and commercially viable energy technology, this is not the case for small, decentralized hydroelectric turbines (up to ca. 5 MW). This technology is interesting not only because of its unused potential in Europe, but also for its possible application in developing countries. Turbine designs that do not require a reservoir are of special interest. In this case also, time-staggered feed-in payments are an obvious way of supporting the technology.

Biomass as a transport fuel
Because of the importance of biomass as a regenerative energy source (see chap. 5), the production of electricity and heat will probably not be enough to absorb the entire volume produced by a decentralized system. There will likely be surpluses, at least on the regional level. The production of transport fuel (gaseous or liquid) can be an additional sales channel. If a very positive energy and efficiency balance can

be demonstrated (as for waste wood, for instance), sales should be guaranteed by a partial fuel tax exemption. As the agricultural industry can purchase tax-free energy anyway, the support could be easily managed by restructuring such subsidies. Supporting this technology is important for the (tropical) developing countries also.

Photovoltaic energy

While the use of solar energy for heat production is hardly in need of support anymore – apart, perhaps, from investment in the qualification of companies that can install the relevant systems –, photovoltaic technology is still at the very beginnings. New surface preparation processes and materials are promising considerable efficiency improvements. In this situation, it is even more important to strengthen the demand, which would also lead to the further development of complementary infrastructures e.g. for installation and maintenance, and to gain a better command of the manufacturing technologies involved in series production. Hence, the demand should be stimulated through state purchase programs. An example of such a program would be to have a photovoltaic system installed in one in every three public buildings within the coming 5–7 years (the figures were chosen so that each installation of a solar energy system in this period would coincide with major roof repairs, the construction of a new building or a basic modernization). This kind of support appears more effective than feed-in payments, which have to be of very considerable amounts at this stage of the technology, in order to have any effect. Independently, the funds for basic research should be increased significantly, especially for research on solar energy (see section 7.2).

Technology development for "interrupted" regenerative energy sources

An additional problem with regenerative energy sources is their often changeable availability. Such fluctuations are most pronounced for wind energy, but they also apply to decentralized biomass refined through microturbines, which, in more northern regions, is only available during certain seasons. Intelligent system innovations are needed in this field, in order to buffer these fluctuations. Apart from direct storage options, which presently suffer from excessive efficiency losses (e.g. batteries, night storage), the grid as a "buffer" is a conceivable solution ("summer sun from the Mediterranean, winter wind from Scandinavia"). The first demonstration projects for a grid management of this kind are already running in Spain, Scotland and California. However, this type of technology development must be put on a much broader basis and has to be tested for specific regional conditions.

Another difficulty is the integration of regenerative energy sources into existing energy consumption processes, with the aim of substituting fossil energy carriers. Here, too, we find a multitude of (industry-)specific adaptation problems (e.g. that heat cannot be supplied at the current temperature level), and the optimal combination of different energy carriers is often an issue. The situation has to be avoided where a specific investment needs to be effected for each individual carrier (e.g. when using different gas qualities). The technological development in this field is still in its very early stages and should be supported – also according to the considerations presented in section 7.3.1 – from public funds. As is the case with all other proposals, specific support measures depend on which development point the technology has reached on the learning curve: Initial demonstration and dedicated pilot products should be supported with public R+D grants, while market introduction and self-commitments for rapid diffusion require subsidies.

Implementing the results of basic research

To achieve a lasting progress of technology development, a broad spectrum of basic research is needed in the EU (the reasons are given in chap. 7.2). Equally important, however, is to bring the possible applications of such research projects into the equation, not least for deciding which technologies should be supported further. For this decision can only be made, in many cases, when semi-technical systems or pilot plants are already in operation. Considering the enormous task of converting about 50% of the energy supplied in the EU to regenerative sources within 50 years, there must not be any "technology dumps". To avoid this, we must create a seamless innovation chain, from basic research to market penetration.

7.8.3
Excursus: Can we choose between different learning curves? – Outlines of a theory

Introduction

Learning curves appear in the economics and business literature since long. In economics the most cited references are Matthews (1949–50) in the trade literature and Arrow (1962) in the growth literature. In the business literature Boston's "Perspectives on experience" (1972) has generated a big push. In empirical research it has become a habit to estimate learning curves for products and processes. Once the product or process to be considered in empirical research is determined this is a straightforward exercise. However, when firms decide which variant of a product or process should be produced or used, they implicitly decide – among other things - on the choice of a learning curve. An interesting example is the case of wind energy. Here the German and Swedish firms did choose to start with large types and the Danish firms did choose for small types. This looks – at least at the first glimpse - like a choice of learning curves, a phenomenon not yet discussed in the economic literature. However, some doubt that there is really a choice. We present the arguments of both views of this seemingly new problem.[7]

The defence of a 'unique learning curve' view

From an *ex-post* view one could say that those who have chosen the large types have not gone through a learning curve process and have learned nothing. They could not do so because the working of their prototypes was too bad. If nothing works, you cannot identify weaknesses and therefore you cannot learn. Moreover, some have argued already at the time of decisions making, i.e. is *ex-ante*, that it could be known that larges types would not work and that they were the wrong starting point. As all this depends on subsidies for research and development this view holds that there was a pre-emption of the learning process for strategic reasons of some firms

---

[7] We do not discuss the argument – sometimes heard from the business press – that firms starting with large types did want to show that these technologies were not viable.

involved. Ultimately this view boils down to saying that there was no choice of learning curves but rather there was only one alternative, namely a beginning with small types implying a unique learning curve for wind energy. In other words, when there is only a unique learning curve, the choice of a wrong starting point – for whatever reasons – prevents learning.

If this view is true then an important conclusion may be that it is important to ensure competition when learning processes, which need subsidies, are started up. In the present context competition came from foreign countries with firms using different concepts.

The defence of a 'choice of learning curve' view

In defence of the 'choice of learning curve' view one could argue that small firms, starting in a garage or similar small size buildings, are forced to choose small types, whereas for large firms only large types are economically interesting and they would never try to compete with small-type pioneers. By implication, small firms start with learning curves of low cost levels per machine whereas large firms start with learning curves having high costs per machine. As a choice this can only occur if both types are technically feasible. If they are, self-selection makes sure that large firms choose large types and small firms choose small types and each chooses the corresponding learning curve. Moreover, if they are technologically feasible but not economically profitable development subsidies may be necessary. Therefore another important aspect is that the Danish subsidized the small types whereas Sweden and Germany in the first instance subsidized the large types. In this sense self-selection by firms was complemented by political selection. However, an open question at this moment is whether or not Danish (Swedish or German) subsidies were available for large (small) types.

If two or more curves are chosen, an interesting question is, which ones leads to faster or longer learning. Each of the learning processes may be stopped for a great diversity of reasons: lack of financial support, technical difficulties or slow learning inducing pessimistic expectations or other aspects.

Conclusion

Whether a choice of learning curves is possible depends on its technological feasibility. Whether it occurs also depends on economic circumstances. Both views presented may be true in some cases but not in others. Therefore there is no need to try to exclude one of the views. What we get from this note, however, is the consideration of a new aspect of technology choice.

8 Implementation Problems of a Sustainable Innovation Strategy

8.1
Actors in the "sustainability arena"

In focusing on the actors, we acknowledge the reality of modern, pluralistic democracies in Europe. No longer is the state a remote entity, detached from any interests, a neutral keeper and preserver of the common good; in wide areas, it is now a "player", a moderator, an implementer and conveyor of whatever emerges as policy in the "parallelogram of social forces". The exception is the state monopoly on the use of force (internal and external security). The "non-players" are left, as voters, to support or reject entire "packages" (election manifestos and leadership teams), without the power to influence individual decisions.

The actors (or the "elites", as they are called more candidly in countries that can look back on a longer democratic tradition) are distinguished by certain possibilities of influencing (within rules prescribed by law and culture) the "result" of the game. Associations can mobilize organized voters' groups, which are critical for the outcome of an election ("swing votes"), if they perceive their interests to be threatened. Farmers' associations are considered particularly effective in all western democracies, because the subsidies they receive are still rising in most countries, in spite of their dwindling numbers. Other actors draw their power from "setting agendas", for instance the media, which not so much determine *what* the citizens think, but what they think *about*. The scientific community and the NGOs (non-governmental organizations) belong on this list, too. Finally, the various actors enjoy different degrees of economic power: The more open the economic area, the more politicians and bureaucrats have to fear losing mobile production factors (and knowledge and capital are much more mobile than labor).

In the following, we will analyze the important actors in the implementation of sustainable energy innovations.

8.2
The attractiveness of sustainability goals from the perspective of selected actor groups

For each group of actors, we first aim to define their direct economic interests and identify the behavior that results when the actors want to increase their individual utility. However, for any group of actors, the utility calculation does not lead to a decision automatism. It is rather the case that all actors can choose from a certain range of actions, which they can use for a more or less sustainable behavior. They can also act as a "homo politicus" (Petersen et al. 2000) and engage as such for the common good, or as "political entrepreneurs" to help change the framework conditions. In every group of actors, such homines politici can be found, albeit often in a marginal role and, hence, only of limited use as partners in reform coalitions.

Voters (citizens): The economic theory of voting distinguishes between two motives for taking part in an election:

- Consumption motive: Independent of the outcome of an election, voting always provides some utility, for instance because the voter acts in harmony with his civic awareness, or as an expression of an enjoyable political engagement. The importance of the consumer motive depends crucially on the level of education and information. For the consumerist voter, the contents (sustainability) are less important than the feeling of being part of an (intellectually) attractive struggle.
- Investment motive: Whether a voter takes part in an election, and how he casts his vote, depends on the utility that can be gained by voting (cost-benefit ratio). But does a vote for a party that campaigns for sustainability improve the individual benefit (income, well-being)?
  - This question is only relevant if the voter can detect any differences in the election programs, concerning sustainability. Important factors for this perception are the level of education on the part of the voter and the accessibility of the issue to a high-impact public presentation ("environmental scandals"). The latter is hardly achievable, given the dryness of the topic (only 10–15% of the German population know the term "sustainability").
  - In addition to this, the expected effects of a sustainability policy need to be assessed. In principle, such a policy improves the situation of future generations, and it usually demands sacrifices from the present population (e.g. concerning the use of resources). , but Only for some voters or voter groups, the policy is of advantage already today (e.g. because it promises job security for workers in the solar technology industry). Therefore, the majority of today's electorate can hardly be expected to vote for it.

Even if an innovation-oriented sustainability policy does not involve socio-economic restrictions (sacrifices) on the whole, there will be some groups of losers. Compensation payments can help in deterring them from active resistance, at least, and make them tolerate some reforms. It can be difficult to bridge the time-lag between implementing reforms, the reinforcement and reorientation of the innovative dynamics triggered by the reforms, and the emergence of overall economic returns which are necessary to finance compensation for innovation losers . One possibility of bridging the time-lag would be through (temporarily) increased public debt (advance financing of investments in the future). However, because of the debt burden already accumulated, this instrument would find little approval among voters. To make the investment aspect attractive for voters, one must find an approach to sustainability policy whose positive effects are visible and perceptible as quickly as possible, whereas its negative effects are hardly noticeable. This condition is met e.g. by establishing new institutions (sustainability council, sustainability reporting), whose costs are widely spread and which still can have a considerable (publicity) effect and, hence, can trigger or enhance the dynamics of sustainability (change of path). In summary: Investments in information ("popularization") and education can increase the approval rate for sustainability goals. "Unnoticeable" institutional reforms will provoke little resistance, but still set the course.

Politicians (political parties): Politicians do not try to implement the preferences of the voters, primarily, but pursue their own objectives (income, status, power, or asserting an "ideology"). To achieve these goals, politicians need to be (re-)elected and to accumulate a sufficient number of votes to this end. Hence, the thinking and actions of politicians is dominated by election cycles (which is, at least, longer than the quarterly time horizon of managers); any long-term policy will find it difficult to gain the attention of people holding political power. Sustainability goals offer little attraction for competing politicians, since future generations simply do not vote in the next election. Long-term benefit is often outweighed by a perceptible burden on today's voters. Still, sustainability might be relevant in vote swaps (logrolling), i.e. in agreements between different minorities for the purpose of securing a joint majority. In this area, sustainability can be very attractive indeed, if the aim is to integrate economic, social and ecological objectives. Thus, even today's minority interests (e.g. ecological positions) have a chance of prevailing.

Public bureaucracies: Because of an information asymmetry, the administrators instructed by politicians to implement political decisions have some latitude in their decision-making and can use this for pursuing their self-interest (a form of the principal-agent problem). Not threatened by unemployment and enjoying few income incentives, the self-interest lies predominantly in increasing their available budgets and, thereby, their capacities for co-determination, and their access to status symbols (official cars, office space etc.). This creates inefficiencies such as excessive costs of service provision and an expansion of services beyond the optimal volume. The latter can have positive effects for sustainable development, once environmental agencies have been established. The self-interests of the bureaucracy are furthered by close links with the political class (high proportion of civil servants in parliaments). However, in a federal state or an economic area like the EU, the momentum of the bureaucracy is restricted by competition between regional bodies (possible migration of enterprises). In ministerial bureaucracies, the appointment of political henchmen can soften the principal-agent problem. In this way, the influence of (ministerial) bureaucracy on the formulation of the political agenda can be reduced as well. In regulatory agencies, a "capture problem" has been observed: With time through the regulation period, a peaceful coexistence develops and mutates the regulatory agency from a supervisory body to an advocate of the regulated sector, leading to a counterinnovative atmosphere. Another obstacle to innovation is the risk aversion of regulating and licensing agencies (concerning e.g. the licensing of new facilities; it took about ten years, for instance, to establish the rules for emissions from biogas systems, which are different from fossil-fueled plants). There is a factual preference for old, known risks over new (possibly lower) ones.

Special interest groups (NGOs etc.): Special interest groups try to move politicians and bureaucrats to decisions that benefit their members, i.e. to gain additional profits for them (rent seeking). Interest groups that have a small number of members with homogeneous preferences are the easiest to organize. In general, it is easier to organize and to assert / prevail established interests (defense of "acquis") than interests that would profit from innovations. The advantages of the latter are often broadly spread and powerful organizations still have to be formed (asymmetric degree of organization). The influence of special interest groups thus favors the status quo; at least, it causes delays in innovation and hinders long-term reforms. Confronted with conflicting demands from special interest groups, the politician looks for the political-economic optimum (Blankart 1998, p. 498 ff.): He will promote environmental policy measures, for instance, only to the point where the electoral gains from environmentally aware voters are just compensated by the drain of votes that might be caused e.g. by employment arguments. Apart from votes, party donations are another important factor determining the optimum.

Among the NGOs, *environmental organizations* play a special role because they declare themselves explicitly as advocates for future generations and their right to (essential) resources, which we would have to preserve. In the debate on sustainability, environmental organizations fulfill an important function as "pacemakers" and an "ethos-shaping force" (*Sachverständigenrat für Umweltfragen* 1994, Tz. 388). They can also be seen as an institution that reduces transaction costs: Since the citizens cannot rely on immediate perceptions concerning environmental issues, and the clash of the experts often leaves them at a loss, they increasingly depend for their judgments on intermediary institutions, which have built for themselves a capital of trust and whose statements, therefore, carry some credibility. Thus they can contribute to the diffusion of sustainable innovations. Some companies are already tapping this potential by entering co-operations that go beyond eco-sponsoring and touch on the core business. Boosted by media attention, the NGOs have gained an influence far more important than their budgets and staff numbers would suggest, which sometimes provokes questions concerning the legitimacy of their actions. Still, internal conflicts about the assessment of energy innovations (e.g. wind energy) occasionally lead to the political paralysis of this group of actors. Within the scope of their specific competencies, the relevant NGOs should contribute to the implementation of energy innovations and involve themselves in more co-operations (also see section 8.4).

Consumers: Through their buying behavior (including boycotts), consumers can instigate lasting progress for learning and innovation processes in enterprises ("demand pull"). Thus, under conditions of competition, they have a decisive influence on the sustainability of products and processes. First of all, consumers are out to maximize their individual utility; they do not ask whether a cheap product is supported by environmental destruction or child labor. But if sufficient information is made available, a certain preparedness to pay for sustainability aspects can be mobilized. State-sponsored and private labels (such as the *Blaue Umweltengel* or *Bioland* in Germany) can help reduce the costs of information. Consumer associations, consumer protection groups and independent product testers, too, can encourage sustainable consumption patterns and be supported in this role with public funds. Since education and culture determine consumption patterns, the school system can also contribute to a transformation in consumer behavior. There are, however, limits both in principal and of a practical nature (paternalism, the attempt of the state to define preferences, or overburdening the school system with all sorts of "extra" services). On the whole, nobody has really yet succeeded in changing the preferences of consumers, apart from "pioneer groups", in favor of sustainable products, be it cars or green electricity. "Sustainability" is hardly noticed among the general flood of information, or it is drowned under the massive advertising for non-sustainable consumption.

In addition to this, consumers' interests are generally difficult to organize. Powerful organizations have only emerged for some specific aspects, e.g. automobile or tenants' associations. Private households can influence the direction of the economy and the innovation process in their role as *savers,* too (in Germany e.g. annual private household saving is about 120 billion € money assets are about 7.5 trillion €). Sustainable investment – the keywords are green money, ethical investment and SRI (Socially Responsible Investing) – becomes increasingly important. Nevertheless, the vast majority of private households still let the savings banks, life insurances and building societies choose the criteria for, and contents of, their investment decisions, which is why investment funds with their demands for shareholder value are still increasing their influence on enterprises.

Enterprises: As the buyers become more prepared to pay for sustainability, the enterprises will turn their attention to it, take a close look at their product portfolio and production processes, and re-orientate them accordingly. The question is whether enterprises in a competitive environment, particularly those faced with the short-term demands of shareholders, are limited to this reactive role, or if there is scope for pro-active sustainable action (Kurz 1997). Some examples do suggest that a corporate leadership oriented to ecological and social sustainability does not necessarily exclude economic success. In other words: Sustainable corporate policies beyond ecological market niches are possible even under the present conditions. Such pioneering sustainable enterprises exploit innovation potentials by

- accepting limitations concerning (short-term) returns, which they hope to be compensated by their improved long-term strategic position in their markets,
- opening niche markets by offering additional ecological and social services, which are seen to be met with a higher willingness to pay,
- entering co-operations with other groups of actors (stakeholders) for projects like *Greenfreeze* and *Green-TV.*

In this respect, dealing with sustainability-related ideas becomes an indicator of the degree of modernization achieved by an enterprise. Pioneers and niche suppliers play an important role as forerunners for a wider diffusion. Since these are SMEs, in most cases, it is probably most important to strengthen them by creating framework conditions that favor start-up and medium-sized enterprises. However, the competitive market environment sets limits on the ecological pioneering role of an enterprise. Here we need the "political entrepreneur", who supports legislation that would force his competitors to act in an equally environment-friendly way. "This is a task, predominantly, of the industry associations" (*Sachverständigenrat für Umweltfragen* 1994, Tz. 125). To focus the attention of managers on sustainability issues, the enterprises can be obliged to undertake certain reporting duties (energy balance sheets, sustainability reports etc.; see also the instrument of the eco-audit (ISO 14000 and EMAS)). In many companies, energy costs are still hidden among general expenses and thus escape the attention of managers. More transparency in this matter, and setting energy saving targets in service-sector companies too, would create the pressure that sensitizes for innovative ideas.

Large enterprises are often considered the only actors that can really make a difference. These same corporations, however, also find themselves under increased justification pressure. Profits have ceased to be a sufficient legitimation. Large enterprises are required to carry responsibilities in the regional context, e.g. for respecting cultural identities; this applies all along the value chain. In this respect, they are quite vulnerable and depend on the security offered by networks (stakeholder co-operation, most notably). It is in their self-interest to be part of the sustainability discussion. But "enterprises" are an extremely heterogeneous actor group. Interests diverge strongly, depending on the sector (polluter sector vs. environmental protection industry/ services) and the size of the company (SME vs. global players).

The energy system is also divided into various segments (mobility, heating, electricity) with, in some cases, diverging interests (substitutional competition), but also with considerable links and overlaps. The interests of the *electricity industry* have changed fundamentally since its deregulation (in Germany since 1998). Under conditions of regulation, the energy suppliers were able to shift the costs of environmental laws and some pro-active initiatives (energy-saving advice, give-away energy-saving light bulbs) to the customer, simply by raising electricity prices, so that there was no need for resolute resistance against energy-saving demands. In a competitive market, this off-loading of costs is more difficult and resistance will probably harden. The increasing insecurity of the market strengthens the interest in flexible facilities with shorter amortization periods. Hence, the resistance against more decentralized structures with smaller units could dwindle, in the longer term, once the problem of "stranded investments" has been defused. One should also note the changed role of the power-engineering sector (power plants, CHP plants), from where technology-push effects can originate. Deregulation has led to the breakthrough of product innovations. From one homogeneous good, electricity, different qualities of electricity have developed, including sustainable options (natural electricity, solar electricity etc.).

This differentiation within the sector also creates starting points for new (innovative) coalitions of actors (Greenpeace, BUND as electricity suppliers, see chap. 7.6.2). The innovation process could be hindered by increasing monopolization tendencies in the sector. The *Wissenschaftliche Beirat Globale Umweltveränderungen* ["scientific advisory board on global changes in the environment"] demands: "As large, financially powerful actors, the energy supply corporations must be co-opted as partners while, at the same time, binding framework conditions for their actions must be agreed" (WBGU 2001, p. 9). The rethinking towards efficient energy services – Thomas Edison's old dream: sell lighting, not electrons – is a slow process (even slower in the US than in Europe). Especially among the ex-monopolists of the grid-dependent energies, relying on the security of the established technology is still more common than an orientation towards innovation (e.g. through energy performance contracting).

Employees and trade unions: Employees and trade unions are interested, above all, in securing their income (growth), on the basis of preserving their jobs. The latter is often interpreted as the preservation of concrete, existing jobs (in coal mining, the construction industry etc.) instead of the overall employment level in an economy. This results in resistance against structural change, which would entail higher demands on regional mobility and on job mobility. To gain the employees' and trade unions' approval for innovation and long-term policies, we first need a sufficient safety net for the losers and provisions for further education and training for getting out of the loser position. There is plenty of evidence that the interests of the individual trade unions are varied and that trade unions, too, can develop a strong interest in sustainability. Initially, the trade unions had left the responsibility for the various aspects of sustainable development (economic growth and employment, social security, warding off ecological hazards) to the state, but now they see themselves, increasingly, as an actor group, co-shaping and co-operating in the wider economy and in companies, thus shouldering part of the responsibility (Hildebrandt and Schmidt 1997, p. 186).

Science: There have always been high hopes that science could play a leading role. However, these can hardly be fulfilled, for several reasons: The differentiation of science into well-defined disciplines is often in the way of dealing with issues that can only be approached in an interdisciplinary manner. Secondly, one should remember the great uncertainty of our knowledge concerning the relevant issues (see section 3.1). And finally, science cannot decide on the objectives that should be pursued by states or organizations; science can only, but importantly enough, make transparent alternative means-end relationships. Public policy can raise the interest of the science system in sustainability by displaying a demand (granting research funds). In this way, it determines what will be a research topic and can thus create counterbalances to the commercial interest in short-term profits. Plurality in the scientific consultation and decision-making bodies can soften the dominance of certain paradigms and their exploitation for special interests.

Conclusion: On the whole, the institutional arrangement outlined above does not produce enough incentives for sustainable energy innovations. However, and this applies to all groups of actors, there are margins for action that can be employed for a (more) sustainable behavior. If this happens, a self-amplifying reform process can set in. The burden on state action, which remains indispensable, will be eased. Sustainability requires and allows that citizens, as members of various actor groups, participate in the reform process. The question is how to invigorate self-organization and learning processes. The strength of an innovation-oriented sustainability policy lies in the prospect of win-win constellations, meaning that more ecological sustainability is possible without any restrictions on socio-economic goals. However, this only holds for the longer term and on balance, there are no quick (social) returns. We are dealing with a long-term policy to bridge the desynchronization between costs and returns. First of all, rearrangements in public budgets are required, primarily new priorities in subsidy policy, and reforms of the framework conditions, so that individual actor groups do not lose out and hence are not moved to organized resistance e.g. by forming resistance coalitions. For politicians, innovations only become interesting, from the votes perspective, when public opinion clearly demands them, and when more or less "ready-to-use" solutions become discernible, only waiting for their wide application through diffusion. Advance action to improve the conditions for basic innovations, on the other hand, is much more difficult to implement.

8.3
Instruments and their attraction from the perspective of selected actor groups

The choice of instruments is not a purely "technocratic", objectively, scientifically determined step; it rather involves value judgments and distribution issues. Each instrument has specific side effects. Choosing an instrument is always a decision about the distribution of burdens and benefits and, therefore, entails a value judgment. This limits the ability of science to offer "objective" statements and, at the same time, mobilizes special interest groups. These groups are active not only in the definition of political goals; they also try to influence the choice of instruments (because of the distribution effects). This also applies to an innovation policy aimed at sustainability. In the following, we will consider the effects of innovation policy and various environmental-policy instruments.

Innovation policy: The logic of the political process has a formative influence on the choice of the instruments of an innovation-oriented sustainability policy. Generally, one has to consider the following facts:

– Votes are won more easily with structure-preserving measures than with an innovation policy, whose widely distributed benefits will only emerge in the future.
– Politicians are interested in quick success (within the current election period) of "their" policies, but any innovation policy must be designed for the long term.

This leads to a general public underinvestment in innovation policy, compared to structure-preserving measures, although the political system also offers considerable incentives for innovation policy, at least for certain variants:

– R&D subsidies can buy the approval of (subsidized) enterprises and employees, with only a limited burden on the general taxpayer. This instrument is also in the interest of bureaucracies, for which it opens up the wide field of grants allocation and project monitoring.
– Specific support for (major) projects is particularly visible and hence attractive for politicians. However, support funds must also flow to small and medium-sized enterprises, because these are numerous (and represent a large pool of votes). Existing institutions and procedures (research bureaucracy) still suffer inertia and develop their own momentum.
– Basic research, especially in science and technology, is publicly financed not least because of pressure from the private sector, which receives free services and qualified scientific staff from it.

These mechanisms ensure that public innovation policy and support will not disappear from the political agenda. On the other hand, when budget cuts are necessary, innovation policy is particularly threatened, because it yields positive effects (apart from the direct effects of government spending) only in the long term.

Because of the constellation of interests outlined so far, the implementation of an efficient innovation policy is by no means guaranteed. Providing subsidies is much less attractive, politically, than showing an engagement for regulatory reforms (of tax and competition laws, for instance). To a certain extent, the reluctance of politics concerning the latter area can be explained by the fact that there is not enough reliable knowledge on necessary regulatory reforms.

International location competition can have a salutary effect concerning the priority given to innovation policies and their efficiency. This allows those disadvantaged by the status quo to bring a more credible threat (migration abroad) into negotiations and thus helps them to more negotiating power. It can advance reforms of the framework conditions, but it can also lead to a subsidies race and waste of funds on a global scale. These unwelcome effects can be avoided by international agreements that put limits on subsidies (e.g. on EU or WTO level).

Environmental policy measures: Another important determinant of sustainable innovation activity is the choice of environmental policy instruments. In the following, we will point out the counter-innovative distortions to be expected in the political process.

(1) Regulation

- Especially in the presence of imminent or acute danger, politicians look to highly visible instruments so that voters take notice of their actions. Regulation (e.g. in the form of prohibitions) meets this requirement.
- Regulation opens up margins of discretion for administrations. Thus, it obliges the aspiration of bureaucracies to expand their activities (and budgets).
- Regulation holds some attraction for enterprises, too. The latter maintain their negotiation margins with environment protection agencies and, because of the "conservative" bias of regulatory law, they encounter hardly any challenge concerning their ability to innovate: Any "state of the art", which might be stipulated, has a long life time – not least because industry does not see any incentive for improvements in such conditions (the "conspiracy of silence of the chief engineers").

(2) Taxation: Eco-taxes are hampered by the problem that politicians tend to be held responsible for the obvious costs (new taxes) rather than for the benefits (ecological and fiscal relief effects). Neither their allocation effect (ecological goal) nor the revenues they yield (fiscal goal) is easy to predict, which is why they are difficult to sell even to sympathetic sections of the electorate. Once the allocation effect sets in, some sectors will contract (perceptible negative employment effects) and resistance coalitions of business leaders and trade unions will emerge. Distribution effects, which tend to be regressive, affecting the vast majority of the electorate.

(3) Tradable permits: Certificate solutions are difficult to communicate; the connection between positive effects and the political initiator is hardly discernible. The ecological effectiveness should make this instrument attractive especially for environmental organizations; in reality, however, they reject it, often vehemently, as "selling (out) the environment" and point to control problems. Enterprises – with the exception of eco-innovative and shrinking companies – fear a rigid reduction regime and are concerned about the unpredictable prices of certificates. This instrument is most likely to be accepted if it increases the flexibility of a regulatory framework (bubble concepts, joint implementation).

(4) Subsidies (direct public expenditure and planned revenue losses through tax allowances): The same allocation effect as described for the tax instrument can be

achieved through subsidies, albeit with different employment and distribution effects (also see section 7.2.2). Concerning the chances for enforcing such a policy, one has to distinguish two cases:

- Case 1: The subsidies (in support of sustainability) are paid for from the general budget and thus by the entire tax-paying population (leading to an increased state and tax share). In this case, enterprises in polluter sectors, as well as their work forces and customers, are better off and will more likely approve these measures than those of an eco-tax regime. Environmentalists are satisfied because their ecological goal is achieved. Taxpayers might put up resistance against the increased tax burden, although they can hardly attribute it directly to the subsidy, or might not notice it at all.
- Case 2: The subsidies (in support of sustainability) are financed through cuts or cancellation of other (non-sustainable) subsidies, aiming at a constant state and tax share.

Generally, case 1 will probably offer better chances for implementation. The ever more intense international tax competition, however, increasingly limits the application even of this politically "optimal" approach. For the foreseeable future, debt-financed sustainability subsidies will hardly be accepted in conditions of massive public debts, even if they are factually equal to an investment in the future. Aside from this, there is still the EU stability pact with its deficit ceiling of 3%.

(5) Voluntary agreements: Politicians find voluntary agreements attractive, especially when there is a marked information deficit (information asymmetries) and a regulatory solution would confront them with considerable difficulties such as high transaction costs, lengthy legislation procedures and the necessity of European or international harmonization. Governments and administrations are interested in voluntary agreements because they are negotiated directly, bypassing the parliament, and thus bring more competencies to the executive. The administration, for instance, are given more tasks and funds for monitoring such agreements. The enterprises are attracted to them because voluntary agreements usually involve agreeing on goals less ambitious than the ones originally tabled, and they are binding only to a certain, rather low degree. Interests turning against a flood of self-commitments are difficult to organize, because this flood would (only) damage the common good: the quality of the environment (delays in the achievement of environmental protection targets) and the free and democratic constitution of the society (risk of corporatism, which can be contained, however, through the requirement of parliamentary approval).

(6) Information instruments (reducing the information costs for the consumer, improving market transparency, informed preferences): Different kinds of marks and labels are of special importance here (see chap. 7.6.3). Private labels emerge from the market process (as "internal institutions") and are subject to only very limited state regulation (concerning their abuse; in Germany, for instance, through the Act against Unfair Competition, *UWG*). However, such eco-, social or sustainability labels have become so popular among enterprises that they cannot contribute to market transparency anymore. This label "inflation" also affects state-regulated labels *(Blauer Engel* etc.). Regarding enforceability, the following applies in general: The lower the level of requirements, the easier it will be to get the approval of

the companies/sectors affected, and the more likely the environmental/consumer protection organizations will object to it because the requirements are watered down too much or the labels are misleading. The higher the level of requirements, the more resistance will arise among the enterprises and the more approval the labeling system will receive from environmental/consumer organizations.

(7) Provisional result: In the political process, instruments such as labels, self-commitments and subsidies have a much better chance of implementation than regulations; eco-taxes and certificate solutions are least likely to be implemented. For each instrument, chances for implementation and acceptance can be improved through modifications (exceptional rules, compensation etc.). Such modifications, though, are usually at the expense of the efficiency or the effectiveness of the instrument. If the structure of the instrument mix leads to the credibility of the goal being questioned, the innovation effect is undermined because the change in expectations and, consequently, the fundamental reorientation of search and discovery processes concerning basic innovations will not succeed in this case. The interests of the relevant actor groups favor the implementation of instruments that are primarily suited for assisting the (more) rapid diffusion of innovations.

Hence, considering the existing constellation of actor groups, the set of measures presented in the preceding chapter offers a good chance of implementation, compared to other options. Yet, the working group sees possibilities to further improve implementation chances and to create better framework conditions for sustainable energy innovations.

8.4
Starting points for improving the chances for successful implementation

If public policy is not trusted and charged with the sole responsibility for transforming the framework conditions, one has to ask what government can do, at least, to activate the social self-organization potentials. These cannot replace mandatory decisions, but they can initiate them, pave the way and thus improve their acceptance.

(1) *Goal formulation and commitment:* Visions and binding targets have two functions: motivation and control. Their mobilizing effect can be used to maintain the pragmatic process of gradual reform (piecemeal social engineering). Targets, too, are an indispensable element of a continuous improvement process. Sustainability goals can be formulated in Local Agenda 21 processes as well as in national sustainability programs. The CO_2 reduction target set by the Federal Government of Germany (25% between 1990 and 2005) developed a considerable effect. The EU Commission is trying to achieve a commitment by formulating a variety of targets, e.g. increasing the share of renewable energy resources in the total energy consumed from 6% to 12% (1997 to 2010), and from 13.9% to 22.1% in the generation of electrical power.

(2) *Co-operation:* In addition to markets (contracts) and hierarchies (directives), co-operation is the third form of coordinating individual plans. Co-operation can avoid the high transaction costs of market co-ordination and the counter-innovative effect

of hierarchical solutions, provided there is (developed) a certain mutual trust. The readiness and ability to co-operate are part of the social assets of any society. Hence, their preservation and improvement is not only the means, but also the end of a sustainability policy. Such co-operation can emerge as a result of learning processes, or it develops from the necessity of finding a common solution with those who hold "veto positions". It can also be furthered exogenously, by state action, state support for co-operation or the threat of regulatory substitutes for co-operation. Anti-trust law can be an impediment to enterprises co-operating with each other. Cartel agreements between companies can help an innovation prevail, but they still represent a partial restriction of competition. Under the threat of state regulations, arrangements such as the *Duales System Deutschland* (DSD), which shows characteristics of a monopoly, came into existence. Statutory regulations concerning car recycling encouraged car manufacturers, suppliers and disposal services to enter new forms of co-operation, with the aim of maintaining a "large degree of competitive neutrality in the disposal process" (Krcal 2000, p. 5); but even so, a possible form of competing for customers, in the area of innovative disposal solutions, is eliminated in this way.

(3) *Capacity building:* By building and embedding (in terms of organization, financing and communication) new capacities in politics (e.g. through parliament committees and "Enquête commissions"), administration (in government departments and agencies), science (e.g. research institutes and new university courses) etc., the chances of a successful implementation of sustainable innovations can be structurally improved. This would make unnecessary the continuous renegotiation of certain decisions, procedures and resources. By linking up existing institutions and creating new ones, actors are placed in new contexts and new actors are established, whose purpose is to help sustainable energy innovations prevail and to maintain continuity.

To overcome the "short-termism" of democratic politics, there are numerous suggestions for creating supplementary institutions, which, by virtue of their independence from government and/or because of their long terms of office, should be free of the pressures of day-to-day politics. For instance, there could be a "Sustainability Agency", endowed by parliament with the necessary instruments for realizing concrete sustainability goals, which it would apply at its discretion.

Germany can look back on mostly positive experience with largely independent institutions such as the *Bundesbank*, the *Bundeskartellamt* [Federal Anti-Trust Agency] and the auditing authorities (Rechnungshöfe). However, all these institutions have their clearly defined tasks, and their effectiveness can be measured by quantitative standards (e.g. price stability or market concentrations). Their actions are based on statutory rules, which can always be changed by a majority in parliament. Sustainability, on the other hand, touches almost every field of politics and each aspect of life. Therefore, the formation of competencies must focus on specific areas – or one would create "councils" outside the political process.

Instead of trying to consider all aspects of sustainability for the design of such an institution, we should restrict ourselves – relating to the terminological approach presented in chapter 2 – to individual, well-defined areas. The focus of this study, – sustainable energy innovations, –can be reflected, accordingly, in the institutional framework. The clearly defined object area would be the energy sector, with a view to sustainable innovations. The annual reduction of CO_2 intensity by 4%, which is taken as a robust assumption in this study, represents a clear orientation mark. Hence, establishing e.g. an institution like an Energy Efficiency Agency (EEA) would be worth considering. It would popularize the efficiency revolution, report on its progress (monitoring function) and identify barriers to innovation. To facilitate approval for its establishment, such an institution would not be given any further competencies, initially. In Germany, it could be developed by expanding the German Energy Agency (dena).

But before one creates new institutions, one should investigate whether existing potentials can be exploited for the objectives identified. If there is no institution able to undertake the respective task on its own, one should try to succeed through linking up existing institutions. In the area of policy consulting, existing consulting bodies should be networked more effectively and urged to issue joint recommendations for action concerning concrete, precisely formulated problems, which need to be approached in an interdisciplinary manner, as is the case for sustainable energy innovations. (In Germany, such institutions could be e.g. the *Sachverständigenrat zur Begutachtung der gesamtwirtschaftlichen Entwicklung* ["council of economic experts"], the *Rat von Sachverständigen für Umweltfragen* ["council of environmental experts"] and the *Wissenschaftlicher Beirat Globale Umweltveränderungen* ["scientific advisory board on global changes in the environment"]).

Thus emerges a two-stage procedure: Firstly, one has to examine which tasks can be undertaken by existing institutions. As the second step, one must identify gaps or create new institutions, if new competencies are necessary. This bottom-up approach to institutional reforms is more productive and efficient, in our opinion, than the top-down establishment of new institutions for new tasks, institutions that are not bound to certain tasks, but are only oriented to interdisciplinary conceptions.

Beyond designing institutions, the political process has produced a multitude of new forms of shaping political intentions, most notably in recent years: citizens' action groups, round tables, citizens' forums, workshops on the future, sectorial consensus dialogues and Local Agenda 21 initiatives. Only the social search process will show which of these forms is to endure. Perhaps the most important contribution of all these bodies and institutions is that they promote common learning processes and loosen up rigid frontlines and defensive attitudes especially against innovation. Although this involves enormous efforts, and much of it does not (seemingly) lead to any result, the contribution of such informal groups to the development of a new communication culture must not be underestimated. With regard to sustainability issues, in particular, we have learned that the competent discussion must not be left to the experts and political-professionals alone.

(4) The chances for a successful implementation of an innovation-oriented sustainability policy can be increased by devising the right *instrument mix*. In this way, resistance can be broken down and coalitions can be formed. Compensation (payments) to loser groups or powerful counter-coalitions (reform coalitions) can improve the chances of a successful implementation. Compensation payments are possible, to a very limited degree, through social security, unemployment and social benefits, adaptation subsidies, retraining and mobility subsidies, while sales and

job guarantees are out of the question. Still, even these "defensive" policy areas can contribute significantly to the acceptance of an innovation strategy.

The strategy proposal of the working group for sustainable energy innovation aims at an instrument mix that is optimized towards implementation. Subsidies in the introduction phase and voluntary agreements and information instruments in the diffusion phase, as opposed to e.g. an eco-tax, are unlikely to lead to the formation of massive counter-coalitions.

(5) *International diffusion of institutional innovations* (also see Jänicke 2000): Nations are also in competition with each other. This competition is about attractive framework conditions or, more precisely, the most attractive combination of framework conditions. Therefore, successful institutional solutions (an independent environment agency, tax or certificate solutions, voluntary agreements, national sustainability plans etc.) are imitated and thereby extended internationally. Unilateral initiatives and pioneering positions, and with them the competitive advantages and disadvantages of enterprises in the pioneering country, will be eroded in this way. Yet, there are no clear criteria for the success of an institutional change. The impact on votes, cost-benefit balances, eco-balance sheets etc. cannot be established beyond doubt, which is why it is difficult to predict which change will find imitators. Nevertheless, it has to be asserted that a sustainable innovation policy can have positive sustainability effects both directly (through process and product innovations) and indirectly (through the international diffusion of institutional innovations).

8.5
Conclusions and perspectives: an Alliance for Sustainable Energy Innovations

If all actor groups independently follow their own rationality, there is little chance of a successful implementation of an innovation-oriented sustainability policy:

– Although sustainable development, as a vision (Leitbild), finds broad approval, political majorities for the (binding) formulation of concrete sustainability targets are very difficult to organize.
– Even if this succeeds, the political process leads to a situation where the most efficient instruments for achieving sustainable development goals are not employed, i.e. there are distortions of the institutional innovation process.
– Sustainability policy and innovation policy are part of a long-term policy, which, in the political process, is subject to the same "laws" that affect every such policy. It finds itself confronted with the dominant attitude of protecting any accumulated assets, rights and entitlements and the orientation to quick successes. However, the crucial difference is that innovation policy is supported by interest groups that are better equipped to prevail. Therefore, an *innovation-oriented* sustainability policy, on the whole, has better chances of prevailing. It makes possible new coalitions of actors including, especially, innovation-oriented enterprises.

The existing institutional design is not the invisible hand that turns action guided by self-interest into an advancement of the common good. Instead, the political process leads to a Prisoner's Dilemma situation with suboptimal results for both actors. The dilemma situation can be resolved by

– changing the goals structure and interests constellation for individual actors (which can happen, for instance, if a sustainable-oriented form of the homo politicus emerges) or
– co-operative behavior (resulting from learning processes and repeated game playing).

To hope for fundamental changes in the goals structure and interests constellation, or for any action against the individual or group-specific rationality, would be unrealistic and unpolitical. However, even under the given framework conditions, all actors have some latitude concerning their behavior, which they can use in favor of sustainable innovations. Within every actor group, there are forces for overcoming resistance and barriers against innovation, for instance in the enterprise sector:

– "Revolutionaries" (sustainability pioneers) in every enterprise (in middle management, among the workforce, etc.),
– the (subversive) innovative power of the "outsiders" in a sector (e.g. the producers of renewable energies),
– a technology push coming from equipment manufacturers, or
– extra-sectorial substitutional competition (e.g. communications technology and facility management).

These forces help change counter-innovative framework conditions set by authorities; they aim at creating an environment in which they can achieve better results. Such interests and a certain preparedness to act as a "political entrepreneur", i.e. to engage for a reform of the framework conditions, also exist in the case of sustainable innovation. Still, because these interests are difficult to organize, it can take a long time for them to prevail over the established interests.

In a climate of essential openness and common principles, visions and threat perceptions, co-operative solutions emerge spontaneously. Every actor group can take the initiative and a (temporary) leading role in this process. What is needed is co-operation in changing coalitions of actors. What could such reform coalitions look like?

– Consumers are natural allies in the innovation process, since their desire for differentiation at least opens up chances for niche markets. The difficulty lies in the next step: making sustainable solutions attractive for mass consumption.
– Environmental organizations can be a driving force, but their influence is minor and, presently, still in decline. Furthermore, their ability to act is limited due to positions critical of technology. Conversely, hoping for an "efficiency revolution" may be a new form of the naïve belief in progress, and the forced acceleration of the innovation process may work against a necessary deceleration (Entschleunigung).
– Science: Almost every specialist discipline has accumulated human capital that works in solving sustainability problems through innovation. There is a particularly strong, but not very powerful interest in, and potential for, an innovation-oriented sustainability policy in this field.
– Entrepreneurs: Even under the present unfavorable framework conditions, some entrepreneurs have managed to successfully exploit innovation potentials. Such

pioneering suppliers play an important role in the initiation of a wider diffusion. They are interested in institutional reforms, but they are difficult to organize, as they are scattered over all sectors. There are numerous examples of networking on the enterprise level (stakeholder approach), whereas co-operations embracing the entire economy and aiming at institutional innovation are the exception (e.g. the eco-tax campaign of the BUND in co-operation with some German corporations in 1997).

The removal of impediments to co-operation and the explicit support of co-operative solutions through public policy would be an approach to reforming the institutional design that could be implemented without much resistance. The *Institut für Organisationskommunikation* cites the following examples of successful state support for co-operation (IFOK (1997), p. 127): The *Deutsche Bundesstiftung Umwelt* supporting the environmental centers of the crafts associations, the *Bioregio* project of the BMBF, the quality alliances in various sectors and the Heidelberg climate pact. From what are initially, relatively informal, voluntary and non-hierarchical co-operations (networks), new institutions may develop – or dissolve the co-operations again.

Hence, if the interest constellation of all actor groups allows an engagement for sustainable energy innovations, we have to ask how one could organize this potential in a political effective way. How can the lack of awareness of the energy problem be remedied, the priorities shifted, inertness factors removed and the fragmentation among the actors with a partial interest in sustainable energy innovations be overcome? This would require a powerful initiative, an *Alliance for Sustainable Energy Innovation*, whose principal goals would have to be:

– increasing the public awareness of the divergence between the energy demand and the demands placed on the energy system (see the set of goals given in chapter 3); promoting an engagement of sustainable innovations (publicity through dedicated events, coordinated presence at trade fairs and conferences etc.),
– identifying obstacles (especially inefficiencies) against sustainable energy innovations; establishing new solutions, and
– practical support for sustainable innovation, e.g. through information, documentation (best practice), consulting (especially influencing company boards), mediating concrete, problem-related co-operations (transfer center).

Such an alliance would bring together various actors:

– Enterprises and their associations (e.g. in the solar energy sector or environment-oriented energy services, and "traditional" energy companies),
– environmental organizations with a practical interest in implementation and an emphasis on the energy sector,
– scientific institutions researching or developing sustainable technologies,
– energy agencies and institutions for the promotion of technology,
– consulting and service companies with innovative ideas.

The success and returns of such an alliance flow from the efficiency pool, whose productivity does not decrease, but rather increases with the number of participants – a typical network effect (hence we are not dealing with a "pool", but with a continuously replenished supply).

The starting point for its implementation could be a conference of high-ranking delegates, which formulates a "mission statement" based on the finding that, due to the divergence of energy consumption trends and central demands of protecting the environment/resources, an acceptable standard of living, concerning energy supplies, is not guaranteed for the future. On this basis, a foundation could be established, whose capital would be raised jointly by the state and by private contributors. Earnings of the foundation could finance the infrastructure required for the network of an alliance for sustainable energy innovations. State participation in the foundation is justified and necessary because of the social importance of the issues involved, but must not lead to state dominance. The core objective is to mobilize state initiative and private, civic-society initiative. Comprehensive participation is certainly not intended; the objective is rather the efficient organization of previously fragmented interests. The alliance must neither be a mere think tank nor an informal PR club (like another Factor 10 Club); it must rather be designed to follow a professional approach. To begin with, national solutions are conceivable. For the medium-term, however, a Europe-wide alliance, or at least a close link-up between national networks, should be the aim. We cannot elaborate on the detailed organization of such an alliance (for instance in working groups or sections according to sectors of the economy or different technologies), but the responsibility for the first step should be clearly placed with government and with politicians, who would have to bring together the high-level actors.

9 Responsibility for the "Energy Hunger" of the Developing Countries – How Sustainable Energy Innovations Can Help

9.1
Basic considerations

Although the measures proposed here must be judged on a global scale, they largely focus on the EU. This was not only necessary for limiting the complexity of the study; it was also the obvious procedure because democratically legitimized sustainability goals for the energy sector have been formulated on the European level, and sustainable energy innovations can play a significant role in achieving these goals. Still, our study would be incomplete if we failed to examine the possible effects of the strategy recommended here on the countries that consume very little energy at present, but which will "catch up" quickly in the future. Hence, the discussion of the role of sustainable energy innovations in the developing countries, which we present in this final chapter, takes the form of an outlook. In terms of the normative component of the sustainability concept, we are dealing with *intra*generational justice here, not with *inter*generational justice. Leveling the energy consumption gradient between different countries, which is documented in chapter 4, by way of a catch-up race cannot possibly be sustainable, because all the progress achieved in the industrial countries could easily be outweighed in this way.

About two billion people living in the developing countries do not have access to commercial energy. Many of these people are in want of food, drinking water, local healthcare and education. They mainly depend on firewood as an energy carrier, with human and animal waste as additional sources. This kind of energy use shows grave effects in some regions: deforestation, soil erosion and desertification. The daily effort for the procurement of fuel and water restricts the time that humans, especially women and children, could use for other productive activities. Moreover, the use of "primitive" stoves or fireplaces for cooking and heating leads to smoke emissions and thus to considerable health problems. Without access to modern forms of energy, reducing poverty through economic and social progress will hardly be possible. In total the absence of opportunities for self-improvement in underdeveloped rural areas is regarded as the structural cause for the migration into cities:

> The greater availability of commercial energy in cities is one of the driving forces behind urban migration. Supplying energy at reasonable prices to rural areas could contribute, along with other factors, to a decline in rates of urbanization (EC and UNDP 1999, p. 3).

Although the poorest countries in the world contribute little to the emission of greenhouse gases, they will be the main victims of the impending climate changes. The list of effects of global warming published by the IPCC includes shrinking agricultural yields in most tropical and subtropical regions, less available water especially in subtropical countries and a growing risk of flooding in many parts of the world (IPCC 2001b). Compared with industrial countries, developing countries are far less able to soften such effects; they cannot possibly finance expensive coastal protection measures, for instance.

The paradigm of sustainability is based on the two principles of intergenerational and intragenerational justice. In this sense, the developing countries demand equal chances for their economic growth, and equal entitlements concerning emissions. However, this can only be realized if the affluent countries drastically reduce their own emissions and if the expansion of the energy system in poorer countries is kept as "climate-neutral" as possible. The obligations of the industrial countries can partly be met through technology sharing, technological progress and innovation (see section 2.4), but even this is a colossal task. The worldwide shaping of sustainable energy systems, as a whole, is an enormous challenge for development policy. This challenge must be faced by all parties concerned.

The underlying problem, as described in chapter 4, is the population growth and the desired economic development, which will lead to rising energy consumption. Nevertheless, the underdeveloped countries are entitled to aspire to an improved quality of life. To avoid problems with the present form of energy provision (section 4.2), a development based on fossil resources, which was the path taken by the industrial countries, must not be repeated. It would clearly be in the self-interest of the affluent countries to help the underdeveloped regions to develop not only economically, but also ecologically. For the energy sector this means, firstly, it must use energy in intelligent ways and, secondly, it must increasingly use solar resources (see section 4.3.2). Hence, the challenge is to manage the transition to the "clean" technologies that are best adapted to local requirements.

9.2
Reorientation of development co-operation in the energy sector

As far as the task of building sustainable structures is concerned, only modern energy technologies can be considered for the construction and reshaping of energy systems in developing countries, although we must not assume that these technologies can simply be replicated in the target country. Any high-tech solution must be appropriate for and, if necessary, adapted to the human requirements and conditions of the respective region.

It is still the case, predominantly, that the technologies are developed in the industrial countries. Any such development should be judged not only by its potential applications at home, but also by its potential for use in other countries (see sections 5.3 and 7.8). Co-operation can only be credible and successful if a broad energy portfolio is used by the developer nation as well.

The adaptation of systems to varied conditions, and the manufacture of more robust, reliable, cheaper and more mobile systems, require additional R&D efforts. Sparsely populated rural areas with infrastructural deficits are particularly well suited for the employment of decentralized, regenerative energy systems. The absence of cable grids in such systems is a definite cost advantage. Another advantage – very important for developing countries – is the much lower capital commitment required, in comparison to large power plants.

Many bilateral and multilateral organizations work in the area of development co-operation. Several international institutions initiate and support activities concerning renewable energies and energy efficiency: the World Bank, the Global Environment Facility, UNDP, UNIDO, UNEP, IEA, UNESCO, the UN and the FAO.

Over the last decade, we have seen a paradigmatic change in development policy: The emphasis in development co-operation is now on the war against poverty and on sustainable development (OECD: "Shaping the 21st Century", 1996; World Bank: "Comprehensive Development Framework", 1999, and IMF/World Bank: "Poverty Reduction Strategy Papers", 1999).

This means that energy policy is also bound to the goals of sustainable development and fighting poverty. The construction of environmentally compatible energy systems, in particular, can be a crucial contribution to social, ecological and economic development: Firstly, it would reduce the dependence on imported fossil resources; secondly, the creation of a decentralized energy system helps local economic development, since the investments involved lead to the creation of jobs and livelihoods. Access to modern energy must also be regarded as a basic precondition for the fight against poverty, as it provides the necessary conditions for using modern ICT systems and hence for an education system that is able to compete.

To get modern energies to the underdeveloped regions, markets need to be opened up. Technology transfer on its own will not achieve this. Even in an industrial country, structures must be developed before the first wind energy facility can be installed. Just considering the aspect of physical access, which is a matter of meters in industrial countries and of kilometers in developing countries, the investments in the structures can easily outstrip the investments in the project itself. Thus, the faults and deficiencies in the present energy system must be attributed rather to economic and structural deficits than to technical shortfalls. Concerning energy procurement, this means that developing countries need help with the expansion of physical and virtual infrastructures. An appropriate infrastructure is a basic precondition for many other economic and social development chances, too. In contrast to large central projects such as water reservoirs, which are so "popular" in development politics, the infrastructure for a new, regenerative energy system can be developed throughout a country and can be guided by the demands arising from other factors.

Mere access to modern energy technologies is not enough to reach the high level of technological development that these countries need and aspire to. The construction of a ready-to-use power plant including engineers from the country of origin cannot multiply the know-how across the society and economy of the receiver country. As the infrastructure created by the technology has a wider reach in the case of renewable energies, other important capacities in education, administration, financing and trades can be developed in this way, also. The developing countries would gain from a transition from "technology transfer" to "technology sharing".

9.3
Existing initiatives for sustainable energy innovations

World Bank

The World Bank Group[1] is the most important consulting and financing instrument within multilateral development collaboration. In the past, it mostly carried out structural adaptation programs in co-operation with the IMF, with the aim of promoting macroeconomic stability. In the energy sector, these were mainly large projects involving fossil energy sources or hydroelectric power. The social costs and ecological consequences of these measures were hardly taken into account.

In 1999, a strategy paper, "Fuel for Thought" (World Bank 1999a), prepared jointly by the Environment Department, the Department for Energy, Mining and Telecommunications and the International Finance Corporation, signaled a reorientation of their energy policy. The new policy emphasizes the early integration of country-specific, environmentally sound energy strategies in the planning process. It declares six distinct goals:

- Promotion of a more efficient use of energy and substitution of traditional fuels;
- Reduction of the air pollution caused by fuel combustion in large cities;
- Support for the environmentally compatible production of (traditional) energies;
- Reduction of greenhouse gas emissions;
- Support for partners in order to be able to establish framework conditions for energy markets;
- Strengthening the responsibility of the World Bank for the environmental effect of energy projects.

According to first conclusions drawn one year after the implementation of the strategy, sustainability aspects of energy projects were paid more attention, but the results were still unsatisfactory (World Bank 2000).

NGOs criticize the World Bank for supporting too many non-sustainable energy projects. For instance, since the adoption of the Framework Convention on Climate Change, much larger funds went into furthering fossil energy projects than into measures in support of regenerative energy sources and energy efficiency (see e.g. Institute for Policy Studies 1997 and Sustainable Energy & Economic Network 2001).

As is the case in European politics, energy issues must be positioned at a more strategic level in the World Bank, because they are too important for global development to leave them to the specialist level. Projects should focus on regenerative energy sources. However, to reduce the significant inefficiencies in the energy policies of many developing countries (e.g. large subsidies for energy and the focus on major projects), a whole range of measures is necessary in the area of general energy policy.

---

[1]   The World Bank Group includes the International Bank for Reconstruction and Development (referred to as the World Bank), the International Development Association (IDA), the International Finance Corporation (IFC) and the Multilateral Investment Guarantee Agency (MIGA).

Global Environment Facility

The Global Environment Facility (GEF) is the finance mechanism provided by two international agreements: the Convention on Biological Diversity and the Framework Convention on Climate Change. The Environment Facility is managed by the World Bank Group, the UNDP and the UNEP. The GEF is a decentralized organization for environmental policy measures. As a multilateral finance institution, it supports the developing and transition countries in their environmental protection efforts in the fields of biodiversity, climate change, international waters and ozone depletion.

To reduce greenhouse gas emissions, it assists public and private partners with strategic project interventions that should steer investment, management and policy decisions towards "climate-friendly" solutions. Since its establishment in 1991, it has allocated about $ 1 bil. to 240 projects concerning climate protection. In addition to this, it co-financed ca. $ 5 bil. of investment. This makes the GEF the leading multilateral organization supporting the efficient use of energy and regenerative energy sources in developing and transition countries. The support grants are awarded in addition to traditional aid. In this way, the GEF only bears the costs that arise from considering global climate objectives in the implementation of measures.

Twelve operational programs were developed as elements of the GEF strategy, and four of these programs refer to renewable energies:

1. Removing barriers against energy efficiency and energy saving
2. Increasing the market shares of (already competitive) solutions in the field of renewable energies;
3. Commercializing future technologies with medium-term potentials;
4. Promoting sustainable solutions in the transport sector[2].

GEF support is intended to help the development of global, national and local markets for energy-efficient and renewable technologies and initiate their wide diffusion. To achieve market penetration with technologies that are already competitive (or almost competitive), various barriers have to be overcome. Cost aspects, information deficits, imperfect capital markets, a lack of human and institutional capacities, technological risks, market risks, difficulties with the introduction of new concepts and high transaction costs[3] are cited as the main obstacles (more details in Martinot and McDoom 2000). Cost reduction measures are essential if systems that are not yet profitable are to have any chance. With its programs, the GEF aims at accelerating the process of cost degression along learning curves. (A similar approach is described in section 7.6).

The concrete projects are subdivided into nine groups: solar PV building systems and rural energy provision (e.g. small wind turbines, charging batteries through solar energy, and wind-powered water pumps), grid-bound regenerative energies, solar hot water provision, regenerative energy technologies that are not competitive, energy-efficient product manufacturing and markets, energy-efficient

---

[2]   Other than the first three points, this program was introduced in December 2000 and will, therefore, not be discussed in this study.

[3]   One problem is that "major projects" are often treated preferentially, because the transaction costs of smaller projects are particularly high. This applies, especially, to the early stages of the projects. Here, too, learning effects can help reduce the costs.

investments in industry, energy-efficient building standards, efficient local heating provision through biomass or geothermal energy, and substitution of fuels and recycling of energy.

The various programs use different approaches, the most important of which include the creation of capacities in human resources and institutions, i.e. public institutions as well as enterprises, organizations and consumers. The people on site must be enabled to meet the technical, financial and organizational challenges. Enterprises, for instance, need management and specialist competencies for constructing, installing, operating and maintaining facilities.

Some of the projects run by the GEF are concerned with shaping energy policy framework conditions including the development of national energy strategies, market liberalization, energy price reforms, tax incentives and the introduction of product standards. As is the case in industrial countries, market incentives can provide a sustained boost to sales of environmentally compatible technologies.

Another approach is the development of new institutions and financial services, e.g. agencies or communal institutions, which implement programs for efficiency improvements. The establishment of new ways of financing, e.g. through micro-loans, is often an important aspect of rural solar projects, which enables the population to use these technologies. In addition to this, the GEF offers technical and financial assistance for manufacturers, for instance in the shape of technology transfer or marketing support with the aim of developing markets for environmentally compatible products. In many projects, institutions and partners such as manufacturers, distributors, consultants and NGOs are integrated as participants.

The aim of the measures is, apart from directly and indirectly changing the market, to initiate a long-term, sustainable transformation. Because there are no suitable assessment standards to date, the successes of these efforts cannot be quantified yet. Even though considerable funds have been made available, the financing of "climate-friendly" projects through the GEF clearly has its limitations. It is not sufficient, on its own, to induce the necessary transformation while binational and multinational organizations still support energy projects that are distinctly non-sustainable. The competencies of the GEF could be expanded. For instance, it could act as the program sponsor for all UN organizations that run projects concerning sustainable development and innovation in the energy sector.

G8 Task Force

At the G8 summit in Japan in 2000, the heads of state and governments decided in favor of the establishment of the G8 Renewable Energy Task Force. The aim of this project is to identify barriers and recommend measures to promote the use of regenerative energy sources in developing countries. At the subsequent summit in Genoa, the proposals developed since the Japan summit were rejected; no concrete action plan was adopted.

The Task Force considers it possible, through joint action by industrial countries, industry, NGOs and international finance institutions, to supply a billion people in developing countries with regenerative energy within 10 years. Although, initially, a slightly higher expenditure has to be accepted, compared to the provision of fossil energies, this path would be more cost-effective in the long term than perpetuating "business as usual".

The development of new financing options is referred to as an important precondition. Such new options are necessary for removing market barriers, ensuring adequate returns, and to contribute to risk cover. Another essential factor is a shift of subsidies from fossil energy carriers to environmentally compatible sources. Major progress could be achieved if development organizations and finance institutions would be consistent in their acceptance of the sustainable development of the energy sector as an essential element of their activities.

9.4
What can be done by the EU?

Credibility is an important factor for success. If developing countries are asked to take a sustainable path, industrial countries and the EU must lead by example. An overview of energy-relevant EU policies is given in section 6.4 and appendix 3.

As a concrete measure, the EU can support the positive initiatives of the international organizations and give a higher priority to sustainable energy innovations. The EU's own development policy emphasizes the necessity of a sustainable development and a climate-compatible energy policy linked to the reduction of poverty:

> Access to a sustainable energy system is a key element in promoting social and economic development. The provision of energy, primarily through decentralized measures and the exploration of renewable energy sources, is becoming increasingly important (KOM (2000) 212).

This prescription, however, never became part of an independent strategy for the sustainable provision of energy in the context of development policy. In 1998, the expenditure in the energy sector amounted to 5% of the total sum of EU aid. It is neither broken down into individual measures nor evaluated systematically. Support is granted within programs of all kinds and comprises a wide range of measures, e.g. technology transfer projects, projects for improving energy efficiency, for modernizing the energy sector and for demand-side management, projects concerning regenerative energies, combined heat and power production, energy transport, the liberalization of energy markets, creating capacities etc. From 1996 to 1998, nearly half of the external aid in the energy sector was spent on nuclear safety in the successor states of the former Soviet Union (Cox and Chapman 1999, p. 43). As late as in the year 2000, considerable funds were granted for this purpose. In addition to this, the EU financed electricity purchases by the Ukraine to meet the immediate energy demand in that country.

Considering their importance for the developing countries, energy issues should take a more prominent position within the EU development program. The report of the G8 Task Force (see above in section 9.3) could serve as a starting point for this new energy policy. The target to supply a billion people from regenerative energy sources within a decade, in particular, could provide a proper focus for private investment and public aid payments. To get over the key barriers (costs, insufficient human and institutional infrastructures, high investment costs and difficulties in mobilizing capital, weak incentives and misguided policies), the Task Force recommends a whole set of measures for each of these problem areas, addressing the developing nations as well as the OECD countries and international institutions.

As described above, the EU introduced the recommended support for the market diffusion of regenerative energy sources and energy efficiency innovations in the industrial countries, whereas the necessary prioritizing of the sustainable development of energy systems is nowhere to be seen in international co-operation. Nevertheless, the measures to be taken by the industrial countries, as proposed in chapter 7, could easily be put into the development policy context.

The Task Force emphasizes the need for joint action by the industrial countries, the private sector, NGOs and the international finance institutions. Thus, the recommendation is also addressed at the business community. Among other measures, the G8 countries should encourage large enterprises to enter voluntary agreements concerning the purchase of regenerative energies. The importance of joint ventures and technology sharing is also highlighted. The industrial countries are asked to provide incentives in that direction. As the following section will show, such co-operation could well be in the self-interest of enterprises.

9.5
Global enterprises and "Technology Sharing"

For developing countries, technology sharing for the sustainable provision of energy means that economic growth becomes possible without leading to more energy-related CO_2 emissions or to energy import dependencies that cannot possibly be financed. Compared to large power plants, renewable energies have the advantage of a shorter installation period than other energy systems, easier financing and earlier profitability. At the same time, countries do not experience protectionism from the side of enterprises, but profit from their support and the infrastructures and competencies they create. Medium-term, the developing countries are able to improve their trade balances by having to buy less of their energy supplies abroad, which makes it easier for them to find solutions to wider social and economic problems.

Enterprises acting on the international stage have an important function in assisting the developing countries, and they have three good reasons for preparing their business policies for technology sharing: Firstly, they open up structures and potentials for new markets in this way; secondly, they create and make use of effective and cost-efficient capacities; and finally, they support consentaneous sustainability goals and are seen as "good corporate citizens". On the part of the industrial nations, public-private partnership also means a public engagement for capacity building in developing countries. For instance, the *Deutsche Entwicklungsgesellschaft* (DEG), on behalf of the German government, assists enterprises with opening up markets in developing countries in order to exploit natural potentials for renewable energies.

Successful examples of "technology sharing"
Almost every country in South America suffers from energy shortages, although it has an immense natural potential for renewable energies. Because of its high-performing industrial sector, Brazil is considered the gateway to South America, acting as an example for all other countries of the continent. Under a DEG project, the German market leader in the area of wind energy facilities, Enercon, received assis-

tance in building a wind farm in Brazil. This was accompanied by special bilateral support for wind energy in Brazil through trade relations between Germany and Brazil. In the end, Enercon did not export wind energy equipment to Brazil, but built a new production plant in Brazil and turned its attention to the structural measures required. The sluggish sales development in the initial years was compensated by the technology sharing activities of the company. Firstly, Brazilian employees were trained in Germany and German employees in Brazil, with the effect, for instance, that all the executives at the Brazilian branch are Brazilians. Secondly, with the capacities on site, grid-dependent installations could be developed as simple primary applications, while experience was gathered for the more efficient construction of isolated systems. Thirdly, the autonomous subsidiary was able to profit from the boom in wind energy in Europe, where it could export its products.

In combination with hydroelectric technology for storage and regulation, wind energy can be an optimal supplement for the provision of electric energy. Moreover, it is not located at one central site and thus reduces transmission losses at the fringes of the grid. More than 90% of the electric power in Brazil is hydroelectric. Because of the lack of rain in the last five years, the reservoirs are emptying rapidly now. For this reason, drastic energy saving measures have been in force since June 2001 (20% enforced consumer savings per month). However, this situation made the national energy supplier, Petrobras, put out a contract for 1,000 Megawatts, initially, and the Enercon subsidiary profits as the only producer in Brazil.

European enterprises are now making intensive efforts to exploit the technical potentials, from biomass through fuel cells to energy efficiency. The interactions between renewable energies and efficiency potentials have led to the emergence of a broad, interdisciplinary field of research and development, from which not only the renewable energies are profiting, but other areas of the economy and society, also. Wind energy can develop extraordinary leverage, especially when combined with other energy carriers: "Economics of scale" gained in the area of wind energy brought economic stability to the entire sector of renewable energy. Moreover, the motivation for the development and diffusion of these technologies is strengthened by the technical and scientific link to wind energy, considering e.g. stand-alone wind-solar systems, generating of fuel cells through wind energy or simply the stability at the fringes of grids.

Even those among the large European energy suppliers who, frequently, still assume a critical attitude towards wind energy in public, have long been running wind farms as well as solar energy and biomass projects as practical experiments in transition countries. If these operate satisfactorily, the companies involved will be confident that they can stabilize the weak grids in those countries and open up new markets with the new technologies. Shell and BP, the European oil corporations, are now the largest producers, globally, of photovoltaic solar panels. For their developments, these enterprises often rely on co-operation and know-how transfer with small specialist companies, which are quickly brought to the world market in this way, so that developing countries can also profit from their products and services.

The China Energy Technology Program, led by ABB in co-operation with the Alliance for Global Sustainability, is an example of efficiency improvement and cost reduction. A method was developed for the continuous evaluation of the real effects of electricity production, taking into account the energy technologies in use and the environmental effects from the beginning to the end of the energy cycle. China, which is presently building a new infrastructure for electricity production, is likely to profit from this program: ABB has long-term interests in China and consistently supports projects and activities that will benefit the Chinese economy and people.

For the developing countries, renewable and more efficient energy technologies would be a high-tech solution particularly suited to their needs and practicality requirements concerning the supply of energy. Isolated systems are a case in point. Hundreds of thousands of those are operating in the developing countries, but only in the form of diesel grids. So far, this potential has not been exploited to any significant degree with other energy carriers or systems. Due to the geographic structure of the Philippines, most of the electricity grid in that country is powered by diesel engines, which have to be replaced every ten to fifteen years. As fuels have to be bought with foreign currency, the Philippine government now wishes to employ more wind and photovoltaic energy. The suppliers of the energy technology in use so far were from the US, but now the country is looking to Europe. Following discussions between the energy minister and suppliers from Europe, it became clear that developing the country with renewable energies would be particularly worthwhile if a local manufacturing industry were established in the process. Such an industry would provide the capacities for driving and accompanying a gradual and flexible expansion of the energy system.

Socio-economic framework conditions in the form of calculable political prescriptions can support the transformation of the economy and of individual enterprises: Fundamental criticism aimed at the WEC and IEA, which are influenced too much by the interests of the nuclear and fossil energy industry, led to the foundation of the World Council for Renewable Energies. An International Renewable Energy Agency (IRENA) is to be established under the UN, for the promotion of global technology transfer in the area of renewable energies. In relation to the IEA, the purpose of this agency is the identification of various ways of overcoming barriers and accelerating the diffusion of renewable energies. Governments, enterprises, and stakeholders in industrial and developing countries will be brought together for this purpose.

However, the engagement of ABB in the WEC for establishing the "one-gigaton goal of GHG emissions reduction" shows that changes can be effected even within the existing international organizations. The above-mentioned goal stipulates that the projects carried out by enterprises in various developing and transition countries are documented. The projected target may not be directly verifiable, but the project provides insights into the obstacles and chances for the political and economic structures of a large variety of countries, and it demonstrates the medium and long-term utility of entrepreneurial co-operation with developing countries.

Opening the markets can prevent protectionism, provoke innovations and create competition, which will stimulate business. India, for example, has been promoting renewable energies for some years in order to be able to meet her increasing energy demand. As a result, almost every supplier of wind energy plants has a factory in India now. The employees of these factories are trained on site in India and with the parent companies in Germany and Denmark. Furthermore, practical experience with the use of wind energy in the developing country is delivered together with the expertise of scientists and engineers from India, Germany and Denmark. This capacity building not only had the result that most of the factories are run by Indian executives, but also that home-grown Indian enterprises emerged, which produce components and complete wind energy plants. These companies as well as all other production facilities based in India profit from the increasing demand for wind energy facilities in the Far East. Furthermore, Indian wind energy plants are also sold in Europe now.

9.6
Outlook and further research issues

Even without being able to quantify the effects of the innovations described above on an energy system developing towards sustainability, our study clearly shows the following:

- The present energy system is unsustainable. The current trends do not lead to more sustainability, but rather in the opposite direction. Hence, it can hardly be denied that action is necessary.
- Sustainable energy innovations offer considerable potentials both with regard to efficiency improvements in the use of energy and to the development of regenerative energy sources. Still, it is unlikely that these innovations will prevail quickly and widely enough to induce the required effects (4% annual reduction of CO_2 emissions until 2050) – which only supports our assertion of an urgent need for action.
- The existing conflicts of goals, and the blockades they cause, can be overcome through an innovation-oriented i.e. relatively "painless" instrument mix, especially as we could show that a weighty coalition of actors can likely be formed for these recommendations for action. This focus on implementability is necessary because the gap between the factual trend and the trend that would be necessary appears to grow ever wider.

For the same reasons, we believe that further research does not have to produce more "global studies". As the case for an immediate need for action is already very strong – and, probably, cannot be strengthened any further scientifically – the focus should move to the question as to *how* a sustainable energy system can be achieved. Science can certainly contribute a lot more to such action-related knowledge than it does at present, concerning for instance:

- rules and criteria for a decision under conditions of profound uncertainty,
- an empirically better supported, internationally comparative analysis of the effects of instruments and of the instrument mix for a more sustainable energy system as well as conditions for overcoming the factual obstacles against decisions, and
- the transferability of the energy innovations developed in the OECD countries to developing countries; this needs to be expanded in a technology-specific manner based on existing experience, and the results must be incorporated more effectively in development policies.

However, even the most convincing scientific research findings are no substitute for the willingness of the political and economic decision makers to apply these findings.

Appendix 1
The global energy system

A.
Development of the global use of energy

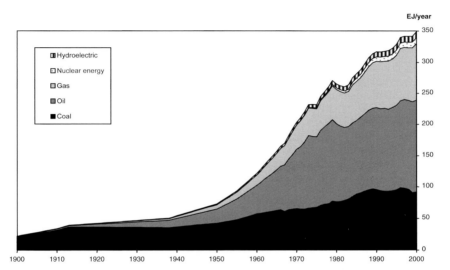

Fig. A.1 Global consumption of commercial primary energies from 1900 to 2000. Nuclear and hydroelectric energy is accounted for as electricity produced. The new, renewable energy sources such as solar, wind and geothermal energy are not included. Their joint contribution totals less than one percent. Data source: Etemad et al. (1991), BP (2001).

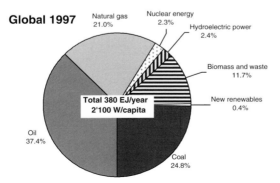

Fig. A.2 Share of energy resources in global energy consumption 1997, including non-commercial energy. Nuclear energy and hydroelectric power are accounted for as electricity produced. Data source: WRI (2001).

B.
Energy production and use in the EU

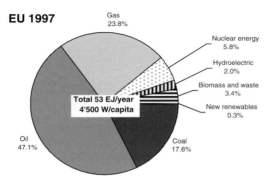

Fig. B.1 Share of energy resources in the energy consumed in the EU in 1997, including non-commercial energy. Nuclear energy and hydroelectric power are accounted for as electricity produced. Data source: WRI (2001).

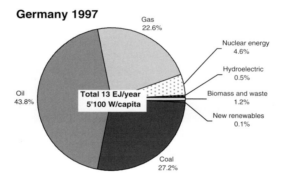

Fig. B.2 Share of energy resources in the energy consumed in Germany in 1997, including non-commercial energy. Nuclear energy and hydroelectric power are accounted for as electricity produced. Data source: WRI (2001).

C.
Energy scenarios

Table C.1 Driving forces of the energy scenarios. The arrows represent the direction and strength of economic growth and the speed of technological development, respectively. The last column shows whether explicit climate protection measures are included in the respective scenarios.

Scenario	Economic growth	Technological development	Climate protection measures
IIASA/WEC A1	↗	↗	No
IIASA/WEC A2	↗	↗	No
IIASA/WEC A3	↗	↗	No
IIASA/WEC B	↗	↗	No
IIASA/WEC C1	↗	↗	Yes
IIASA/WEC C2	↗	↗	Yes

Box 1: IIASA/WEC scenarios

Case A: "High Growth"

Case A is characterized by very high rates of economic and technological development, following the belief that there are no essential limits to human inventiveness. Case A assumes favorable geopolitical conditions and a free market. The economic growth in the OECD countries is 2% per annum and twice this rate in the developing countries. These high growth rates make possible large efficiency improvements and considerable technological progress. Case A comprises three scenarios with different developments in the procurement of energy.

A1: In scenario A1, oil and gas resources are amply available in the future. The two energy resources will continue to be dominant until the end of the 21st century.

A2: Scenario A2 assumes that the oil and gas resources will run short rapidly and that a massive recourse to coal will take place.

A3: In scenario A3, rapid progress in technological developments concerning nuclear energy and renewable energies results in the disappearance of fossil fuels for economic reasons rather then because of their scarcity.

Case B: "Middle Course"

Case B represents a more pragmatic scenario than cases A and C. It starts from modest assumptions concerning the economic growth, technological progress, removal of trade barriers and facilitation of international exchange. Case B would allow the desired development in the southern hemisphere, albeit less uniform and more sluggish than other scenarios. The moderate demand for energy and the slower development of technologies results in more reliance on fossil resources than for all other scenarios except for the coal-intensive scenario A2.

Case C: "Ecologically Driven"

This scenario, like Case A, is optimistic with regard to technological development and geopolitical conditions. However, in contrast to case A it assumes unprecedented progress in international co-operation concerning environmental protection and justice. The future described by it involves a broad spectrum of environmental control technologies and procedures including incentives for the efficient use of energy, "green" taxes, international environmental and economic treaties and technology transfer. There would be a substantial transfer of resources from the industrial to the developing countries. The economic output is smaller than in case A, but larger than in case B, resulting in an equalization of the present economic disparities.

Case C includes measures for the reduction of CO_2 emissions to 2 GtC per annum (a third of today's level) by the year 2100.

In Case C, nuclear energy finds itself at a crossroads. Two distinct scenarios are identified, which both meet the CO_2 targets, but differ with regard to the role of nuclear energy:

C1: A new generation of safe and compact nuclear reactors is developed (100-200 MW capacity in terms of electricity production). This technology finds wide social acceptance, especially in regions where land resources are sparse and which are densely populated, i.e. where the potential of renewable energy resources is limited.

C2: Nuclear energy acts as a transitional energy and will become insignificant by the end of the 21st century.

Source: Nakićenović et al. (1998)

Box 2 – The IPCC scenarios

A1

The story line or family of scenarios A1 describes a future world experiencing very rapid economic growth, the world population peaking around mid-century and declining from then on, and more efficient technologies. The underlying principle is that regions will grow together and social interaction will take place, resulting in a substantial reduction of regional differences between per-capita incomes. The A1 family of scenarios progresses in three groups representing different directions of the technological changes in the energy system. The three groups differ in their technological emphasis: fossil-intensive (*A1FI*), non-fossil energy resources (*A1T*) or a balanced mix of all resources (*A1B*).

A2

Story line A2 describes a very heterogeneous world. The guiding principle is autonomy and preservation of local identities. Fertility patterns among the regions converge very slowly, which leads to a continuous growth of the world population. Economic development is primarily region-oriented and per-capita economic growth and technologies develop in a more fragmented way, and more slowly than in other story lines.

B1

Story line B1 represents a convergent world experiencing a homogeneous population growth peaking around mid century and declining thereafter, as in story line A1. The economic structure, though, changes rapidly towards a service and information economy involving a reduction in material output and the introduction of clean and resource-efficient technologies. The emphasis is on global solutions promoting economic, social and ecological sustainability, including more equality, but excluding additional climate initiatives.

B2

Story line B2 represents a world where the emphasis is on local solutions towards economic, social and ecological sustainability. This is a world experiencing continuous population growth at a smaller rate than in story line A2, a medium rate of economic development and slower, but more diverse technology changes than story lines B1 und A1. This scenario is oriented to environmental protection and social equality, however with an emphasis on the local and regional level.

Source: IPCC (2000)

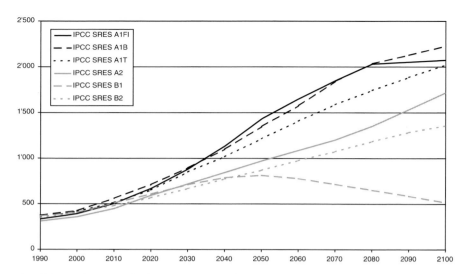

Fig. C.1 Development of primary energy production, in EJ, for the different IPCC scenarios. Data source: IPCC (2000).

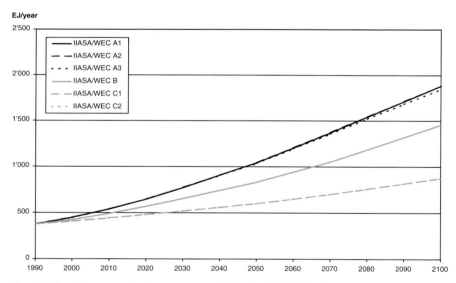

Fig. C.2 Development of primary energy production, in EJ, for the IIASA/WEC scenarios. Data source: Nakićenović et al. (1998).

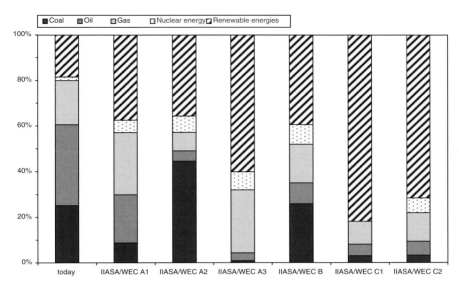

Fig. C.3 Contribution of different energy carriers to global energy production today and in 2100. In scenario IPCC SRES B1, the hatched area represents "non-fossil electricity". Data source: Nakićenović et al. (1998).

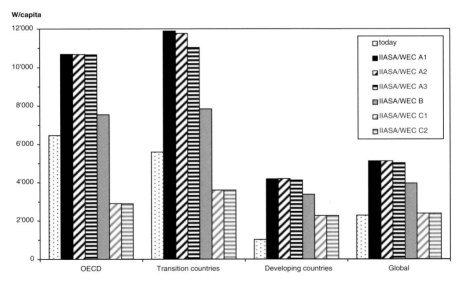

Fig. C.4 Per-capita energy consumption, today and in 2100, of the OECD countries, the transition countries, the developing countries and the whole world. Data source: Nakićenović et al. (1998).

Gt CO$_2$ /year

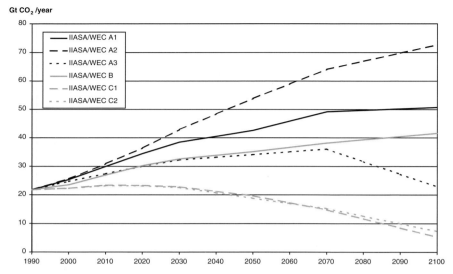

Fig. C.5 CO$_2$ emissions in Gt CO$_2$ per year. Data source: Nakićenović et al. (1998).

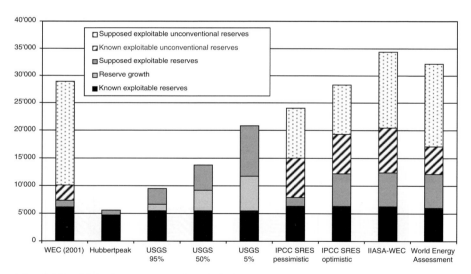

Fig. C.6 Established reserves of mineral oil, in EJ, according to various sources. Data from: Hubbertpeak (2001), IPCC (2000), Nakićenović et al. (1998), UNDP/UNESCO/WEC (2000). USGS (2000), WEC (2001). The figures represented by the rightmost bar (World Energy Assessment) are used for calculating the oil lifespan in table 4.3.

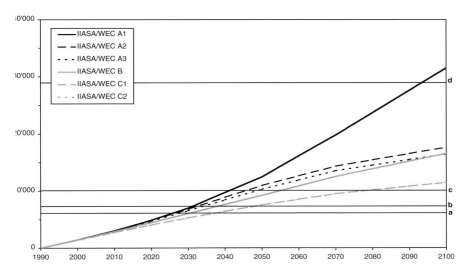

Fig. C.7 Cumulative oil production according to the scenarios and oil reserves in EJ. The horizontal lines a to d represent the reserves according to WEC (2001), see figure C.6: a: known exploitable reserves, b: a + supposed exploitable reserves, c: b + known exploitable unconventional reserves, d: c + supposed exploitable unconventional reserves. Data source: Nakićenović et al. (1998), WEC (2001).

Appendix 2
Unemployment

A.2.1
Elasticity issues in efficiency wage models

In an efficiency wage model of the Solow type, Schneider (1997) points to a condition that determines whether a reduction of labor taxes that is financed through a tax on environmental emissions results in more employment. If the effort, e, of an employee increases with his net wage w/T – where w is the gross wage, 1/T the tax factor and (1-1/T)w the tax revenues per working hour – and with the unemployment rate, and if the production volume x is achieved with the amount of labor L and with the amount of energy-related emissions E, the production function is

$x = f[e(w/T,u)L, E].$ [1]

From cost minimization considerations it follows that the elasticity of the labor input in relation to the wage rate must be equal. From this one can deduce, in turn, that the elasticity of e_u in relation to w must be less than 1 in order to obtain a negative relationship between gross wages and employment. This is expected on the basis of empirical considerations. The total differentiation of e_w (w/T,u)w/e(w/T, u) =1 yields, after rearrangement, $\partial w/\partial u = (e_u - e_{wu}w/T)/(e_{ww}wT)^2$. Ceteris paribus, the relationship between w and u depends on the quantity e_u, i.e. on the strength of the disciplining effect of unemployment on the work effort. This relation is stronger for a higher value of T. In terms of growth rates, the relation can be written down as:

$$\hat{w} = -\beta \hat{u} + \hat{T}$$

Beta is an elasticity coefficient whose value turns out to be important for the result. Assuming a CES production function, the first-order conditions for labor and energy-related emissions yield w/e= p $(eL/E)^{\rho-1}$ (where p is the real energy or emission price and w is the real wage), or in terms of growth rates:

$$\hat{w} - \hat{e} = \hat{p} + (\rho - 1)(\hat{e} + \hat{L} - \hat{E})$$

The wage setting function can be either flat or steep. It can be plotted in the (w, L)-plane (see figure 1) and one can consider the process of a green tax reform. An

---

[1] This function is assumed to be linear-homogeneous with positive, but decreasing marginal products. The partial derivatives are calculated assuming $e_w > 0$, $e_{ww} < 0$, $e_u > 0$, $e_{uu} < 0$, $e_{uw} > 0$.

increase in energy prices pushes up the labor demand. Decreasing energy consumption shifts the labor demand curve downward. A labor tax reduction moves the wage setting curve downward. Scholz showed that the budget implications of a positive relationship between wage taxes and public expenditure necessitates that energy input must fall and unemployment must increase where Beta is of a high value.

From the defining equations for employment, ax=E and bx=L, with a and b as factor input coefficients and the usual definition of the substitution elasticity, $\sigma = -\partial\ln(b/a)/\partial\ln(p/w) = -1/(\rho - 1) > 0$, one obtains the growth rates:

$$\hat{L} = \hat{E} + \sigma(\hat{p} - \hat{w})$$

A comparison with the preceding equation shows that the effort e is kept constant. This too, shows that the volume of employment crucially depends on whether the reduction of employment due to a reduced use of energy is overcompensated by the impact of the factor price changes and substitution possibilities. The factor price changes are interdependent. From the zero-profit condition, it can be derived that:

$$\hat{p} = -\hat{w}\theta_L/\theta_E$$

Fig. 1

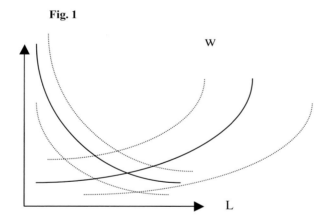

The terms theta represent the cost components due to labor and energy-related emissions, respectively. By substituting this equation in the preceding one and then substituting the wage setting function for the growth rate of w and the definition $\hat{L} = -\varepsilon\hat{u}$, with $\varepsilon = u/(1-u)$, for the growth rate of L, before resolving the resulting equation for the growth rate of u, one obtains:

$$\hat{u} = \frac{-1}{\dfrac{\beta\sigma}{\theta_E} + \varepsilon}\hat{E} + \frac{\sigma/\theta_E}{\varepsilon + \dfrac{\beta\sigma}{\theta_E}}\hat{T}$$

This represents a linearly increasing percentage change of the unemployment rate as a function of the percentage change of the tax factor T. With the energy input and emissions decreasing, this function has a constant positive value on the vertical axis and a negative value on the horizontal axis (see figure 2). This curve is the result of expressing the labor-demand and wage setting functions in terms of growth rates, including the zero-profit condition and the condition of exogenously falling energy emissions. Thus, this represents a consideration of the labor market equilibrium independent of any public budget. Consequently, figure 2 shows the unemployment rate required for any exogenous change in taxation, if the labor market equilibrium is to be secured, or, vice versa, the tax change required for any given rate of unemployment so that the unemployment rate remains compatible with setting the efficiency wage.

Fig. 2

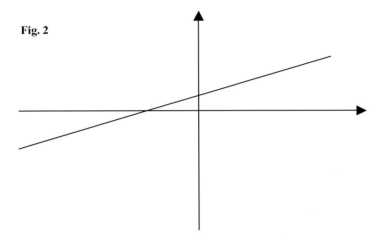

Figure 2 shows three possible constellations. The extreme left side of the graph represents the situation expected by advocates of an energy tax. The rate of change of the unemployment rate is negative, since the impact of the tax reduction is strong enough to encourage low wage settlements. However, if the tax reduction does not have such a strong impact, the result is growing unemployment. Consequently, even if one shifts the wage setting function in figure 1 further towards the bottom right corner, one obtains a weaker labor demand, which is why the demand function must be further towards the bottom left corner. Apparently, the effect of a reduced energy input is dominant here. Even a tax increase (right edge of figure 2) is guaranteed to lead to higher unemployment. At this point, the central question is what rates of change result for taxes and unemployment if public budgets are taken into account. The assumption is constant state expenditure G= (1-1/T)wL + pE. Scholz (1998) considers the reduced form of the complete model and deduces three crucial results:

$$\hat{T}/\hat{G} = \beta_u / DET$$

$$\hat{u}/\hat{p} = \frac{\sigma/s}{DET}(\theta_E/\theta_L)$$

$$\hat{E}/\hat{p} = \frac{\sigma/s}{DET}\alpha$$

where $DET=\{\beta_u\,(1-\tau)\,\theta_L-\varepsilon\theta_E-\varepsilon\tau\theta_L\}/s$ is the determinant of the reduced form of the model, $\alpha=\tau\varepsilon-(1-\tau)\beta_u$, $\tau=1-1/T$, and s the share of state spending in the output x. Starting from there, Scholz shows further that Schneider has not only assumed $\alpha>0$, but also, implicitly, DET < 0. He points out that, in his opinion that is not the usual approach in financial science. It is unclear why this should be the case. The connection between DET and α, which, after all, are composed of the same parameters, also remains unexplained. In the following, we will clarify this connection, in order to demonstrate that the success of a green tax reform crucially depends on the slope of the wage curve, w(u).

An inspection of the definitions of DET and α shows that:

$$DET >(<)\ 0 \text{ if } \beta_u > (<)\frac{\varepsilon\theta_E/\theta_L+\varepsilon\tau}{1-\tau}\ and$$

$$\alpha>(<)0, \text{ if } \beta_u < (>)\frac{\tau\varepsilon}{1-\tau}.$$

The right-hand sides of these inequality relations show that the quotient in the upper condition for β_u is greater than the the quotient in the condition in the second line, because in the former something is added to $\varepsilon\tau$ in the numerator. Consequently, one has to distinguish three cases:

1. $\beta_u > \dfrac{\varepsilon\theta_E/\theta_L+\varepsilon\tau}{1-\tau} > \dfrac{\tau\varepsilon}{1-\tau}$, hence $DET > 0,\ \alpha<0,\ \hat{E}<0, \hat{u}>0, \hat{T}/\hat{G}>0$.

Increasing energy taxes reduces energy emissions, but it also increases unemployment. Out of the three effects considered in connection with figure 1, the reduction in labor demand due to the reduced energy input is the strongest, if the budget context is taken into account. The main reason for this is that a steep wage curve that is shifted downward has little effect on the employment figures. The same is true for energy price increases with a substitution elasticity of less than parity. The additional revenue from an income tax reduction is negative.

2. $\dfrac{\varepsilon\theta_E/\theta_L+\varepsilon\tau}{1-\tau} > \beta_u > \dfrac{\tau\varepsilon}{1-\tau}$, hence $DET < 0,\ \alpha<0,\ \hat{E}>0, \hat{u}<0, \hat{T}/\hat{G}<0$.

An energy (emissions) tax, which leads to a reduction of labor taxes and wages, is good for employment, but it also increases energy consumption and emissions. Out of the three effects expected in connection with figure 1, the one which reduces the labor demand, does not apply at all. The additional revenue from the income tax reduction is positive because the growth of employment, and hence the taxable base, will overcompensate the fall in wages.

3. $\dfrac{\varepsilon\theta_E/\theta_L+\varepsilon\tau}{1-\tau} > \dfrac{\tau\varepsilon}{1-\tau} > \beta_u$, hence $DET < 0,\ \alpha > 0,\ \hat{E} < 0, \hat{u} < 0, \hat{T}/\hat{G} < 0.$

This is the case of a very flat labor supply or wage curve. The employment effect is strong while the wage reduction is minor. As a result, the energy tax discourages the use of energy. Only in this last case is green tax reform successful.

Thus, if a politician wants to know whether he would increase or decrease unemployment by introducing an energy tax, he must know the value of Beta in relation to the other factors. This will be difficult for three reasons:

Firstly, the wage curve features in the model, instead of the usual labor supply function. Based on empirical studies, supply curves are expected to be very steep. Bovenberg (1995) states that a 1%-increase in the wage rate leads to a 0.02%-increase in the labor supply. The result is a nearly vertical function similar to an exogenous labor supply. If the wage setting function enters the model instead of a labor supply function, the question is if this function can look very different from a labor supply function. Independent of the model, the structural equations, which have to be estimated, are always very similar (Pissarides 1998). This would suggest a very high value of Beta and hence an increase in unemployment, which can still be optimal because it would result in an improved environment (Schneider 1997, section IV). Since opinion polls, which are held in such high regard by politicians, show a high priority for employment, it must be doubted if voters and politicians would realize the utility function behind such a result (Böhringer and Vogt 2001). Therefore, it is no surprise that calls for a green tax reform are derived, predominantly, from models involving fixed wages and hence a *horizontal* labor supply curve (see e.g. Nielsen et al. 1995 and Koskela et al. 2001)[2] – and from negotiation models, in which the workforce gets a better deal (see chap 6.2.1).

Secondly, the term $\alpha = -\beta/T+(1-1/T)\varepsilon$ must be positive. In this expression, $1/T$ is the percentage kept by the employee. Graafland and Huizinga (1999) estimate a similar equation, albeit derived from a negotiation model, and obtain semi-elasticities (- $\partial w/w/\partial u$) between 1.5 for the second half of the 1970s and 3.0 for the beginning of the 1990s. To make these comparable with Beta, the semi-elasticity can be multiplied by the unemployment rates at the given time, i.e. the first figure with u=5% and the second with u=10%. This yields values of 0.075 and 0.3 respectively.

---

[2]   Still, in their model, too, an ecological tax reform increases employment only if the tax on the labor input is higher than that on the energy input. If one includes a profits tax in the model, an ecological tax reform increases (decreases) employment if the the profits tax is low (high) (Boeters 2001). The type of model considered here neither includes any interesting labor market contributions nor does it take into account free access to the goods market or any relevant arguments for access barriers.

The higher the unemployment rate, the higher the elasticity. This distinguishes this method from the usual ones, which apply constant elasticities. In figures 3 and 4, two planes are displayed in each case. The *flatter* plane represents the comparative value of Beta, as derived from the Graafland-Huizinga semi-elasticities. The *bent plane* shows, on the vertical axis, the values for the right-hand side of the relation $\beta_u < u(T-1)(1-u) \equiv \beta$ for alternative values of the unemployment rate and the percentage $t=1/T$ of the gross wage kept by the worker. The lower this percentage and the higher the unemployment rate, the larger becomes the value on the right-hand side of the relation. If the elasticity is as high as $\beta_u = 0.3$, the tax factor $t=1/T$ left to the employees must be very small (<40%) for any unemployment rate less than 15% to make possible a double dividend. In the case of a low elasticity, $\beta_u=0.075$, the condition for a double dividend can be fulfilled with plausible values of u and t. Hence, this method does not give an unequivocal answer that would allow a prediction of the change in the unemployment rate.

Thirdly, one can fall back on empirical, general equilibrium models. Böhringer et al. (2001) describe a model of a closed economy with three products, three factors – of which one, energy, is a produced factor – and one wage setting function. For a calibration with $\beta = 0.5$, the authors see unemployment reduced by an energy tax. However, the authors point out that this is achieved by loading part of the tax burden onto the international, immobile and fixed capital stock or its owners. As capital is mobile, this might not work as predicted by the model, which leaves us with the question as to what the results would be in the case of open economies with mobile capital. The arguments of Bovenberg and van der Ploeg, which are presented in the main text of this study, are valid for perfect capital mobility.

Another efficiency wage model – in this case of the Shapiro/Stiglitz (1984) type – goes back to Strand (1996). In the fourth section of his article, taxes on emissions are returned as subsidies for labor costs or output.[3] The first difference to Schneider's model is that the production function, $x = f(N)$, does not depend on a polluting input such as energy. Thus, environmental policy cannot have the negative effect on the labor demand, which, according to Schneider's model, is caused by a reduction in the energy input. The second difference is that the effort z is related to the pollution, $h(N, z)$. Consequently, the reduction of the pollution is an effect of two forces, the change in the employment volume, N, and the increased effort, z: $dh = (\partial h/\partial N)dN + (\partial h/\partial z)dz$. With growing employment, $dN > 0$, the pollution can only decrease if the condition $dh = (\partial h/\partial N)dN + (\partial h/\partial z)dz < 0$, hence $(\partial h/\partial N)dN < - (\partial h/\partial z)dz$ is fulfilled. Dividing both sides by h and multiplying the left (right) side with N/N (z/z) yields:

$$\varepsilon_{hN}\hat{N} < -\varepsilon_{hz}\hat{z}$$

---

[3]   The other sections concern subsidies for the production volume and for the effort made not only for the environment (represented by z in the text above), but also in the production area (y, with $dy = - dz$). As a rule, production subsidies are less efficient with regard to employment (also see Strand 1996, section 5). The conditions remain as restrictive even if the measures in favor of the environment are in competition with the measures concerning output (Strand 1996, section 3).

The percentage change of z and its elasticity in relation to the pollution h must exceed the percentage change of employment and its elasticity in relation to pollution. Otherwise, the pollution can only be reduced by a reduction in the employment volume. This follows already from an inspection of specified functions. If we include the relationship between efficiency wage, profit maximization and benefit maximization through a dual government decision, we obtain the *elasticity conditions* for increasing employment: $-h_{zN}N/h_z > 1$ *and* $\varepsilon_{hN} \geq \varepsilon_{fN}$, which can be interpreted the following way: An increase in employment requires that the elasticity of h_z in relation to N must be more than equal and that the increase in employment boosts the output by a larger degree than it worsens pollution. Thus, two conditions must be fulfilled, if not only pollution, but also employment is to be increased. Again, a politician would have to trust certain estimates – provided there are any estimates that result in the values required for the relevant parameters.

A.2.2
Elasticity issues in negotiation models

Holmlund and Kolm (2000) extended a negotiation model with monopolistic competition and a constant number of enterprises by including a sector of non-tradable goods. The profits are subject to negotiation and the energy is imported. The introduction of an energy tax has no direct effect on employment in the non-tradable goods sector, whereas the labor demand falls in the tradable goods sector. The wage reduction thus triggered boosts employment in the non-tradable goods sector. If the negotiating power and the wages are at exactly the same level in both sectors, the total employment effect is zero, provided the technologies are of the Cobb-Douglas type. If the wages are higher (lower) in the tradable goods sector, the employment effect is positive (negative), since labor migrates to the sector where the trade unions have less negotiating power. In the case of a CES technology with a substitution elasticity below (above) parity, on the other hand, the employment effect is

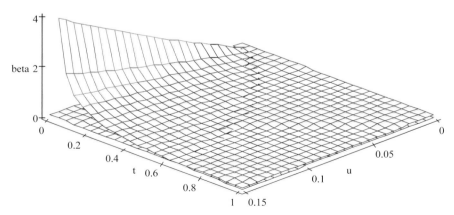

Fig. 3 High levels of unemployment and taxation allow a double dividend if the wage curve has a low elasticity (0.075).

negative (positive) for symmetric sectors, because the wage reduction leads to a slight (heavy) increase in the labor demand in the non-tradable goods sector. Empirical estimates do not allow any clear conclusion concerning the value of the substitution elasticity.

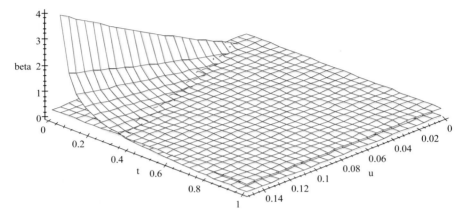

Fig. 4 Only a very high unemployment rate and a very high tax factor allow a double dividend if the wage curve has an elasticity of 0.3.

Appendix 3
Energy-relevant science and technology policies of the European Union – an overview[1]

A.3.1
Importance and integration of sustainability aspects in European energy policies

All three dimensions of a sustainable development are monitored in the context of the annual spring sessions of the European Council, where they are inspected with regard to the extent of their integration in European policies. At the spring session in Barcelona in March 2002, for instance, the European Council examined the progress of the integration of the goals of a sustainable development in the Lisbon Strategy, and assessed the contribution that can be made by the environmental technology sector to the promotion of growth and employment. As a result, the Council recognized the need for further action.

Investments in science and future technologies are thought to be of crucial importance, because without such investments an adaptation towards sustainable developments would have to be achieved, to a larger extent, by changing consumers' behavior. New technologies can be developed through support for innovations. Therefore, the EU and its member states must ensure that the creation of framework conditions that favor innovations (i.e. providing a regulatory push) takes a central role in shaping their policies in order to stimulate the forces that drive innovation: the growth of scientific knowledge (science push) and the growth of demand (market pull).

Public finance support for the technological transformation must focus on basic research (far from the market) and applied research in the area of safe and environmentally sound technologies as well as on benchmarking and demonstration projects in order to accelerate the creation or introduction of new, clean and safe technologies.

In the same context, the European Commission refers to the energy sector, and the promotion of research and technological development (RTD) in this sector, as a central area of present and future policies to implement measures against climate change and increased use of clean, safe and renewable energies. According to the Green Paper published in November 2000, "Towards a European Strategy for the Security of Energy Supply", the objective is to ensure the safe, inexpensive and environmentally compatible procurement of energy while maintaining the economic competitiveness of the European energy market. Against the background of the ca. 50% of the demand currently met by energy imports and the CO_2 reduction

---

[1]   The EU documents discussed in this chapter can be found at http://europa.eu.int/index_de.htm.

requirements arising from the Kyoto Protocol of 1997, the security of energy procurement and its environmental dimension are becoming increasingly important.

A.3.2
Overview of energy-relevant RTD programs of the European Union

The energy-relevant RTD programs of the European Union can be subdivided into
five categories. In the following, we will focus on parts of the first three of those:

1. Energy Framework Program (1998-2002) comprising six specific agendas:
    - ALTENER (Specific Actions for Greater Penetration of Renewable Energy
      Sources),
    - SAVE (Specific Actions for Vigorous Energy Efficiency): Energy effi-
      ciency/savings,
    - ETAP (studies, analyses and prognoses for energy markets),
    - SYNERGY (international co-operation in the area of energy policy),
    - CARNOT (clean technologies for solid fuels),
    - SURE (safety, transport, co-operation in the area of nuclear energy);
2. The European Climate Change Program (ECCP; KOM (2000) 88 fin.)
3. The 6th Framework Program for Research, Technological Development and
   Demonstration Activities (2002-2006);
4. Third state programs: INCO, PHARE, TACIS;
5. Part of the structure programs: e.g. INTERREG, RECHAR.

Framework program for activities in the energy sector (1998-2002)[2], especially:
ALTENER/SAVE
To achieve the strategic objectives of security of procurement, competitiveness and
environmental compatibility, the Commission has formulated Community initiatives, centered on the "Framework Program for Activities in the Energy Sector"
(1998-2002), for optimizing the transparency, coherence and coordination of all
Community activities in the energy sector.

While the RTD programs within the 5th/6th Framework Program for Research,
Technological Development and Demonstration Activities and the support possibilities under the structure programs, e.g. INTERREG, are equipped with considerable
budgets, the "policy programs" ALTENER and SAVE are funded much less lav-
ishly. Among the reasons for this situation is, certainly, that these programs are not
technology-orientated; they are intended to identify the legal, administrative and
institutional impediments to an accelerated market penetration of efficient and
innovative technologies and, ultimately, remove them by political means.

In this way, ALTENER und SAVE are supplemental to the technology programs
of the EU. They start from where technology support programs usually lose their
leverage, i.e. at the point of developing and evaluating activities to remove those
impediments that still hinder the market penetration of technically proven, clean
and efficient technologies.

---

[2]   up-to-date details at http://europa.eu.int/comm/energy/en/pfs_4_en.html.

Consequently, they do not involve investment subsidies in the narrow sense of the term.

- The programs are composed of four parts:
- Legislative measures at the Community level,
- studies to support the work of the European Commission,
- financial support programs for assisting the member states in removing legal and administrative obstacles and information deficits of the relevant target groups, and
- activities in support of information exchange (information networks, databases).

In its approach, the SAVE program addresses the energy demand side (rational use of energy, RUE) and ALTENER the energy supply side (renewable energy sources, RES). This pragmatic dichotomy dissolves as "investments in RES should always be preceded or accompanied by a demand management plan and/or by investing in RUE, since energy from RES should not be wasted through inefficient demand side/end user equipment, appliance and systems"[3]. Therefore, against the background of experience gained in 2001 and 2002, especially such projects that integrate SAVE und ALTENER ("integrated actions on RUE and RES") are to be supported financially.

The European Climate Change program (ECCP)[4]

The ECCP was conceived to include every important interest group in the preparations for joint, coordinated policies and measures to meet the emission reduction requirements arising from the Kyoto Protocol (KOM (2000), 88 fin.). The ECCP focuses on measures in the areas of general issues, energy, transport and industry. The proposed catalogue of measures takes into account, supports and supplements the efforts to integrate environmental issues with other policy areas. "The ECCP also confirms the need to continue research in the areas of climate protection, technological development and innovation" (KOM (2001) 580 fin.). For instance, it recommends emphatically making better use of, and also further developing the existing IPPC (Integrated Pollution Prevention and Control Directive 96/61/EC) guidelines with regard to generic energy-saving technologies and of state of the art applications, continuously updated in technological reference documents[5] and representing IPPC obligations concerning energy savings. Furthermore, it deals with issues of residential and industrial energy consumption (minimum efficiency requirements, energy demand management, promotion of nuclear power plants) as well as with a number of activities in accordance with the White Paper on a European Transport Policy (KOM (2001) 370).

The 5th Framework Program for Research, Technological Development and Demonstration Activities (1998-2002)

One of the four top priorities cited in the 5th Framework Program for Research, Technological Development and Demonstration Activities (1998-2002) is activi-

---

[3]   European Commission, DG Energy & Transport (2002), p. 4.
[4]   Up-to-date details at http://europa.eu.int/comm/environment/climat/eccp.htm.
[5]   BREFs (BAT Reference Documents).

ties concerning "energy, environment and sustainable development (EESD)", which was budgeted with € 2.125 bil. The EESD was again subdivided into the action fields "energy" (€ 1.042 bil.) and "environment and sustainable development" (€ 1.083 bil.).

The 6th Framework Program for Research, Technological Development and Demonstration Activities (2002–2006)[6]

Early in 2000 Philippe Busquin, the EU Commissioner responsible for research, presented a visionary concept for a European Research Area (ERA, see KOM (2000) 6), which highlighted many important areas where the EU is lagging behind its main competitors, the USA and Japan: In the present situation, there are "15 plus 1 research policies" – those of the member states and those of the European Commission –, which often act in parallel and with little coordination. The EC Treaty, on the other hand, provides (in Art. 165) the express possibility that "Member States coordinate their research and technological development activities so as to ensure that national policies and Community policy are mutually consistent". However, this has hardly been put into practice yet (EVA 2001).

The ideas presented with the 6th RTD framework program go far beyond the structure and instrumentation of the 5th Framework Program. The concept of a European Research Area shows an innovative way towards a European "internal market for research". It was thoroughly discussed in the European Parliament, in the Council and also in the European Council of heads of states and governments and met with agreement in principle. The measures under this sixth RTD framework program are carried out in accordance with general objectives such as strengthening the scientific and technological bases of (European) industry and developing its competitiveness.

To facilitate the achievement of these goals, the EC framework program (total budget ca. € 16.27 bil.) is structured around three main priorities:

– Integration of research (€ 13.8 bil.),
– Shaping the European Research Area (€ 2.605 bil), and
– Strengthening the foundations of the European Research Area (€ 320 mil.).

The first point, "integration of research", which is to bring together, and thus pre-structure, the research efforts and activities in seven priority areas, seems to be of particular relevance. Energy-related research and development shall be given "appropriate priority" in this context as, according to the European Council of Gothenburg, global changes, the security of energy procurement, sustainability in matters of transport, the sustainable use of Europe's natural resources and the interactions with human activities are of central interest. The measures to be taken in this priority area aim at boosting the scientific and technological capacities required in Europe for realizing a short and long term sustainable development, including and taking into account the ecological, economic and social dimensions. These activities shall represent a comprehensive contribution to the international efforts of softening or even reversing present negative trends and shall help to preserve the equilibrium of ecosystems.

---

6   Up-to-date details at http://europa.eu.int/comm/research/nfp.html and www.eva.ac.at.

The 6th framework program provides, as a central element, seven "thematic priority areas of research". The contents and instruments are formulated in quite some detail in the so-called "specific programs", where the area "non-nuclear energy" is mainly subsumed in the thematic area "sustainable development, global change and ecosystems". This area consists of the subprograms

– sustainable energy systems,
– sustainable surface transport and
– global change and ecosystems.

It is equipped with a total budget of € 2.120 bil. € 810 mil. is earmarked for the sub-program "sustainable energy systems" for the total life span of the program (4 years), and € 610 mil. for "sustainable surface transport".

A.3.3
The research priorities "Energy" and "Transport" in the 6th RTD framework program

The text of the actual (changed) proposal of the Commission concerning the specific programs (KOM (2002) 43) is presented in the sections "sustainable energy systems" and "sustainable surface transport" below.

A.3.3.1
Sustainable energy systems

The purpose of strategic objectives is to address the reduction of greenhouse gases and pollutant emissions, the security of energy supply, the increased use of renewable energy as well as to achieve an enhanced competitiveness of European industry. Achieving these objectives in the short-term requires a large-scale research effort to encourage the deployment of technologies already under development and to help promote changes in energy demand patterns and consumption behavior by improving energy efficiency and integrating renewable energy into the energy system. The longer-term implementation of sustainable development also requires an important RTD effort to ensure the economically attractive availability of energy, and overcome the potential barriers to adoption of renewable energy sources and new carriers and technologies such as hydrogen and fuel cells that are intrinsically clean.

Research priorities

Research activities having an impact in the short- and medium-term
Community RTD activity is one of the main instruments which can serve to support the implementation of new legislative instruments in the field of energy and to change significantly current unsustainable patterns of development, (which are characterized by growing dependence on imported fossil fuels, continually rising energy demand, increasing congestion of transport systems, and growing CO_2 emissions), by offering new technological solutions which could positively influence consumer/user behavior, especially in the urban environment. The goal is to bring innovative and cost-competitive technological solutions to the market as quickly as possible through demonstration and other research actions aimed at the market,

which involve consumers/users in pilot environments, and which address not only technical but also organizational, institutional, financial and social issues.

Clean energy, in particular renewable energy sources and their integration in the energy system, including storage, distribution and use

The aim is to bring to the market improved renewable energy technologies and to integrate renewable energy into networks and supply chains, for example by supporting stakeholders who are committed to establishing 'sustainable communities' employing a high percentage of renewable energy supplies. Such actions will adopt innovative or improved technical and/or socio-economic approaches to 'green electricity', heat, or bio-fuels and their integration into energy distribution networks or supply chains, including combinations with conventional large-scale energy distribution.

"Research will focus on: the increased cost-effectiveness, performance and reliability of the main new and renewable energy sources; integration of renewable energy and the effective combination of decentralized sources, with cleaner conventional large-scale generation; validation of new concepts for energy storage, distribution and use."

Energy savings and energy-efficiency, including those to be achieved through the use of renewable energy sources

The overall objective is to reduce the demand for energy by 18 % by the year 2010 in order to contribute to meeting the EU's commitments to combat climate change and to improve the security of energy procurement. Research activities will focus in particular on environmentally sound building to generate energy savings and improve environmental quality as well as the quality of life for their occupants. Activities in the area of "polyvalent" energy production will contribute to the Community target of increasing the share of combined heat and power systems (CHP) in EU electricity generation from 9% to 18% by 2010, and will help to improve the efficiency of the combined production of electricity, heating and cooling services, by using new technologies such as fuel cells and integrating renewable energy sources.

"Research will focus on: improving savings and efficiency mainly in the urban context, in particular in buildings, through the optimization and validation of new concepts and technologies, including combined heat and power and district heating/cooling systems; opportunities offered by on-site production and the use of renewable energy to improve energy efficiency in buildings."

Alternative motor fuels

The Commission has set an ambitious target of a 20 % substitution of diesel and gasoline fuels by alternative fuels in the road transport sector by the year 2020. The aim is to improve the security of energy supply through reduced dependence on imported liquid hydrocarbons and to address the problem of greenhouse gas emissions from transport. In line with the Communication on alternative fuels for road transportation, short-term RTD will concentrate on three types of alternative motor fuels that could potentially reach a significant market share: bio-fuels, natural gas and hydrogen.

"Research will focus on: the integration of alternative motor fuels into the transport system, particularly into clean urban transport; the cost-effective and safe production, storage, and distribution (including fueling infrastructure) of alternative motor fuels; the optimal utilization of alternative fuels in new concepts of energy-efficient vehicles; strategies and tools to manage the market transformation process for alternative motor fuels."

Research activities having an impact in the medium- and long-term
The medium- and longer-term objective is to develop new and renewable energy sources, and new energy carriers such as hydrogen, which are both affordable and clean and which can be well integrated in a long-term sustainable energy supply and demand context both for stationary and for transport applications. Furthermore the continuing use of fossil fuels in the foreseeable future requires cost-effective solutions for the disposal of CO_2. The goal is to bring about a further reduction in greenhouse gas emissions beyond the Kyoto deadline of 2010. The future large-scale development of these technologies will depend on a significant improvement in their cost and other aspects of competitiveness against conventional energy sources, within the overall socio-economic and institutional context in which they are deployed.

Fuel cells, including their applications:
These represent an emerging technology which is expected, in the longer term, to replace a large part of the current combustion systems in industry, buildings and road transport, as they offer higher energy-efficiency, lower pollution levels and a potential for lower cost. The long- term cost target is 50 €/kW for road transport and 300 €/kW for high-durability stationary applications and fuel cells/electrolyzers.
    "Research will focus on: cost reduction in fuel cell production and in applications for buildings, transport and decentralized electricity production; advanced materials related to low and high temperature fuel cells for the above applications."

New technologies for energy carriers/transport and storage, in particular hydrogen:
The aim is to develop new concepts for long term sustainable energy supply where hydrogen and clean electricity are seen as major energy carriers. For H2, the means must be developed to ensure its safe use at an equivalent cost to that of conventional fuels. For electricity, decentralized, new and in particular renewable energy resources, must be optimally integrated, within inter-connected European, regional and local distribution networks to provide secure and reliable high quality supply.
    "Research will focus on: Clean, cost-effective production of hydrogen; hydrogen infrastructure including transport, distribution, storage and utilization; for electricity the focus will be on new concepts, for analysis, planning, control and supervision of electricity supply and distribution and on enabling technologies, for storage, interactive transmission and distribution networks."

New and advanced concepts in renewable energy technologies:
    Renewable energy technologies have, in the long-term, the potential to make a large contribution to the world and EU energy supply. The focus will be on technologies with a significant future energy potential and requiring long-term research, by means of actions with high European added value in particular to overcome the major bottleneck of high investment costs, and to make these technologies competitive with conventional fuels.
    "Research will focus on: for photovoltaic: the whole production chain from basic material to the PV system, as well as on the integration of PV in habitat and large-

scale MW-size PV systems for the production of electricity. For biomass, barriers in the biomass supply-use chain will be addressed in the following areas: production, combustion technologies, gasification technologies for electricity and H_2/syngas production and bio-fuels for transport. For other areas the effort will be focused on integrating at European level specific aspects of RTD activities which require long-term research."

Capture and sequestration of CO_2, associated with cleaner fossil fuel plants:
Cost-effective capture and sequestration of CO_2 is essential to include the use of fossil fuels in a sustainable energy supply scenario, reducing costs to the order of € 30 in the medium-term and € 20 or less in the longer-term per metric tonne of CO_2 for capture rates above 90 %.

"Research will focus on: developing holistic approaches to near zero emission fossil fuel based energy conversion systems, low cost CO_2 separation systems, both pre-combustion and post-combustion as well as oxy-fuel and novel concepts: development of safe, cost-efficient and environmentally compatible CO_2 disposal options, in particular geological storage, and exploratory actions for assessing the potential of chemical storage."

A.3.3.2
Sustainable Surface Transport

The White Paper: 'European transport policy for 2010: time to decide' forecasts a transport demand growth by 2010 in the European Union of 38 % for freight and 24 % for passenger transport (base-year 1998). The already congested transport networks will have to absorb the additional traffic, and the trend suggests that the proportion absorbed by the less sustainable modes is likely to grow. The objective is consequently both to fight against congestion and to decelerate or even reverse these trends regarding the modal split by better integrating and rebalancing the different transport modes, improving their safety, performance and efficiency, minimizing their impact on the environment and ensuring the development of a genuinely sustainable European transport system, while supporting European industry's competitiveness in the production and operation of transport means and systems.

Research priorities
The objective is to reduce the contribution of surface transport (rail, road, waterborne) to CO_2 production and other emissions including noise, while increasing safety, comfort, quality, cost-effectiveness and energy-efficiency of vehicles and vessels. Emphasis will be given to clean urban transport and rational use of the car in the city; new technologies and concepts for all surface transport modes (road, rail and waterborne); advanced design and production techniques.

Making surface transport safer, more effective and more competitive: The objectives are to assure the transport of passengers and freight, taking into account transport demand and the need for rebalancing transport modes, while increasing transport safety in line with the 2010 objectives for European transport policy; rebalancing and integrating different transport modes; increasing road, rail and waterborne safety and avoiding traffic congestion.

A.3.4
Specific programs and instruments

The 6th RTD framework program, of which extracts are presented above, is to be implemented through specific programs (KOM (2001) 279 and KOM (2002) 43). Each of these programs is characterized by the type of instruments used in accordance with the objectives and organization of the framework program. The specific program "Integrating and Strengthening the European Research Area" with its two indirect activities, "focusing and integrating Community research" and "strengthening the foundations of the European Research Area" embraces the research and coordination activities.

"Networks of excellence" and "integrated projects" are the new instruments to be applied with priority, where a smooth transition from the traditional instruments to the new ones is to be guaranteed. With the introduction of the new instruments, which were welcomed by the Council and the European Parliament in their resolutions on the European Research Area, the necessity was acknowledged that the forms of Community support for research had to be developed further.

The "integrated projects" instrument is meant to strengthen the competitiveness of the Community or to contribute to the solution of important social problems by mobilizing a critical mass of resources and competencies in research and technological development. Each integrated project is tailored to concrete scientific and technological goals and should produce specific results, e.g. products, processes or services.

"Networks of excellence" are intended to expand the outstanding scientific and technological capacities in Europe by means of a gradual integration of the research capacities that already exist or are emerging on the national or regional level. The objective of each network will be to enhance the level of knowledge in a certain area by building up a certain critical mass of skills. Networks of excellence support co-operation between the outstanding capacities existing at universities, in enterprises (both SMEs and large corporations) and in scientific-technological organizations. These activities, which often cover a number of different disciplines, are oriented to long-term objectives; they are not guided by predefined, specific results in the form of products, processes or services.

A.3.5
Conclusion and outlook

The proposal of the Commission concerning the 6th framework program (2002-2006) is strongly guided by the idea of an ERA. As the most important RTD-relevant activity provided for in the EC Treaty, the 6th framework program will also be the most important instrument for implementing the ERA.

The new framework program is based on the following principles:

– Focusing on a Union-wide approach offering the maximum added value for Europe.
– Devising the various activities so that they have a stronger structuring effect on the research done in Europe; this should be achieved through creating closer

links between the national and regional initiatives as well as with other European initiatives.
– Simplifying and streamlining the rules of implementation by applying the newly defined forms of support and the planned, decentralized methods of administration.

The proposal suggested a total budget of € 17.5 billion Euros, of which 1.23 billion were set aside for the EURATOM component and only 810 million for the non-nuclear energy sector. Thus, the negotiations led to an expansion of the budget compared to the 4th and 5th research framework programs, albeit on a very low level (see figure 1 and chapter 6.4):

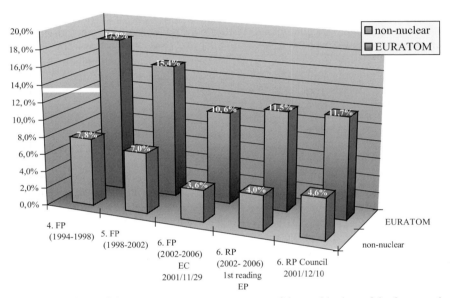

Fig. 1 Budget share of the energy sector as a percentage of the total budget of the framework program

Source: EVA (2001), p. 12; own illustration

The commission assumed rapid progress in its decision process concerning the specific programs and expected, following the adoption of the framework programs on June 3, 2002, (8800/02 (press 132), 2,431st session of the Council), the adoption of the action programs concerning the specific (implementation) programs and a first deployment of funds by December 2002.

Bibliography

AEG (2001) Grünbuch 2000. Nürnberg

Althammer W, Buchholz W (1999) Distorting Environmental Taxes: The Role of Market Structure. In: Jahrbücher für Nationalökonomie und Statistik 219/3+4, p 257–270

Arrow KJ (1962) The Economic Implications of Learning by Doing. In: Review of Economic Studies, Vol 29, p 155–173

Ashford N (2000) An Innovation Based Strategy for a Sustainable Environment. In: Hemmelskamp J, Rennings K, Leone F (eds) Innovation-Oriented Environmental Regulation - Theoretical Approaches and Empirical Analysis. ZEW Economic Studies, Band 10. Heidelberg: Physica, p 67–107

Atkinson G, Dubourg R, Hamilton K, Munasinghe M, Pearce D und Young C (1997) Measuring Sustainable Development. Macroeconomics and the Environment. Cheltenham: Elgar

Bach S, Bork C, Kohlhaas M, Lutz C, Meyer B, Praetorius B, Welsch H (2001) Die ökologische Steuerreform in Deutschland. Heidelberg: Physica

Barry B (1999) Sustainability and Intergenerational Justice. In: Dobson A (ed) Fairness and Futurity. Essays on Environmental Sustainability and Social Justice. Oxford: Oxford University Press, p 93–117

Barry B (1978) Circumstances of justice and future generations. In: Sikora RI, Barry B (eds) Obligations to future generations. Philadelphia: Temple University Press, p 204–248

Barry B (1977) Justice between generations. In: Hacker PMS, Raz J (eds) Law, morality and society. Oxford: Clarendon Press, p 268–284

Becher G et al. (1990) Regulierungen und Innovation. Der Einfluss wirtschafts- und gesellschaftspolitischer Rahmenbedingungen auf das Innovationsverhalten von Unternehmen. München

de Beer J (1998) Potential for Industrial Energy-Efficiency Improvement in the Long Term. Ph.D. Thesis, Utrecht University

de Beer J, Worrell E, Blok K (1998a) Future Technologies for Energy-Efficient Iron and Steel Making. In: Annual Review of Energy and Environment, Vol 23, p 123–205

de Beer J, Worrell E, Blok K (1998b) Long-term energy-efficiency improvement in the paper and board industry. In: Energy, the International Journal, Vol 23, p 21–42

Besch H et al. (2000) Strategien und Technologien einer pluralistischen Fern- und Nahwärmeversorgung in einem liberalisierten Energiemarkt unter besonderer Berücksichtigung der Kraft-Wärme-Koppelung und erneuerbarer Energien – Kurzfassung der Studie. Frankfurt/Main

BFS (2001) Bundesamt für Statistik, Bern

Bhagwati J, Srinivasan TN (1983) Lectures on International Trade. Chap.13, Cambridge MA: MIT Press

Birnbacher D (1988) Verantwortung für zukünftige Generationen. Stuttgart: Reclam

Blankart CB (1998) Öffentliche Finanzen in der Demokratie. München: Vahlen

Bleijenberg AN, van Swigchem J (2001) Schone energie: kern van het energiebeleid. Economisch-Statistische Berichten, Jg 86, p 796–799

Bloemers R, Magnani F, Peters M (2001) Paying a green premium. In: The McKinsey Quarterly, No 3, p 15–17

Blok K, de Jager D, Hendriks CA (2001a) Economic Evaluation of Sectoral Objectives for Climate Change – Summary for Policy Makers. European Commission, DG Environment

Blok K, Harmelink M, Bode JW (2001b) Experiences with Long term Agreements on energy-efficiency Improvements in the European Union. ECOFYS Energy and Environment

Blok K, Turkenburg WC, Eichhammer W, Farinelli U, Johansson TB (eds) (1996a) Overview of Energy RD&D. Options for a Sustainable Future. Office for Official Publications of the European Communities, Luxembourg

Blok K, Eichhammer W, Nillson L, Valant P (eds) (1996b) Strategies for energy RD&D in the European Union. JOU2-CT-0280

Blok K, Turkenburg WC, Eichhammer W, Farinelli U, Johannson TB (eds) (1995) Overview of energy RD&D Options for a Sustainable future. JOU2-CT 93-0280, June

Blok K, Alsema EA, van Wijk AJM, Turkenburg WC (1985) The value of storage facilities in a renewable energy system. Proc. of the Sixth EC Photovoltaic Solar Energy Conference. Dordrecht: Reidel, p 337–342

BMBF (Bundesministerium für Bildung und Forschung) (2000) Bundesbericht Forschung 2000. Bonn

Böhringer C, Ruocco A, Wiegard W (2001) Energy Taxes and Employment: A do-it-yourself Simulation Model. ZEW Discussion Paper No.01-21, Berlin

Böhringer C, Vogt C (2001) Internationaler Klimaschutz – nicht mehr als symbolische Politik? ZEW Discussion Paper No. 01-06, Berlin

Bosquet B (2000) Environmental tax reform: does it work? A survey of the empirical evidence. In: Ecological Economics 34, p 19–32

The Boston Consulting Group (1972) Perspectives on experience. Boston: The Boston Consulting Group

Boeters S (2001) Green Tax Reform and Employment: The Interaction of Profit and Factor Taxes. ZEW Discussion Paper No.01-45, April, Berlin

Bovenberg AL (1995) Environmental Taxation and Employment. In: De Economist, Vol. 143, No. 2, p 111–140

Bovenberg AL, van der Ploeg F (1998) Tax Reform, Structural Unemployment and the environment. In: Scandinavian Journal of Economics, Vol 100 (3), p 593–610

BP (2001) Statistical Review of World Energy

Braczyk HJ (Hrsg) (1998) Regional Innovation Systems. London: UCL Press

Bröchler S, Simonis G, Sundermann K (eds) (1999) Handbuch Technikfolgenabschätzung. Berlin: Edition Sigma

Buchanan JM (1969) External Diseconomies. Corrective Taxation and Market Structure. In: American Economic Review, S 174-177

Bullinger HJ (ed) (1994) Technikfolgenabschätzung. Stuttgart: Teubner

BUND/Misereor (ed) (1996) Zukunftsfähiges Deutschland. Ein Beitrag zu einer global nachhaltigen Entwicklung. Studie des Wuppertal Instituts für Klima, Umwelt und Energie. Basel u.a.: Birkhäuser

Bundesumweltministerium (ed) (2000) Erneuerbare Energien und Nachhaltige Entwicklung. Berlin

Capros P et al. (2000) Einfluß der Besteuerung von Brennstoffen auf die Technologieauswahl – Eine Analyse, Anhang 2. In: KOM 769 endgültig, Brüssel

Carraro C, Galeotti M, Gallo M (1996) Environmental taxation and unemployment: Some evidence on the "double dividend hypothesis". In: Europe, Journal of Public Economics, Vol 62, p 141–181

Commission on Global Government (CGG) (1995) Our Global Neighbourhood. Oxford: Oxford University Press

Costanza R (1991) Ecological Economics. The Science and Management of Sustainability. New York: Columbia University Press

Cox A, Chapman J (Overseas Development Institute) (1999) The European Community External Cooperation Programmes. Policies, Management and Distribution. London

Cropper ML, Oates WE (1992) Environmental Economics: A Survey. In: Journal of Economic Literature, Vol 30, No 2, p 675–740

Daly H (1996) Beyond Growth: The Economics of Sustainable Development. Boston: Beacon Press

Daly H, Cobb JB (1989) For the Common Good. Boston: Beacon Press

Dodgson M, Rothwell R (eds) (1996) The Handbook of Industrial Innovation. Cheltenham: Brookfield

Economische Zaken 3 (1997) Regelingen EZ. Energie Investeringsaftrek, 14 februari, p 19

Economische Zaken 3 (2001a) EINP Subsidieregeling Energievoorzieningen in de Nonprofitsector en bijzondere Sectoren. http://www.ez.nl/subs/01342.htm, download, 15-1-01

Economische Zaken 3 (2001b) EIA Energie Investeringsaftrek. http://www.ez.nl/subs/01342.htm, download, 15-1-01

Economist (18.5.2001) The Bush's Energy Plan

Effectiviteit Energiesubsidies (2000) Onderzoek naar de effectiviteit van enkel subsidies en fiscale regelingen in de periode 1988-1999. Von ECOFYS, OCFEB und Vrije Universiteit Amsterdam

Ehle D (1996) Die Einbeziehung des Umweltschutzes in das Europäische Kartellrecht. Köln u.a.: Heymanns

Eizenstat S (1998) Stick with Kyoto. In: Foreign Affairs, Vol. 77 No.3, p 119–121

Endres A, Radke V (1998) Indikatoren einer nachhaltigen Entwicklung. Berlin: Duncker & Humblot

Enquête-Kommission des 13. Deutschen Bundestages "Schutz des Menschen und der Umwelt"(1998) Abschlussbericht. Bundestagsdrucksache Nr. 13/11200 (http://dip.bundestag.de/parfors/parfors.htm)

Environmental Law Network International – ELNI (ed) (1998) Environmental Agreements – The Role and Effect of Environmental Agreements in Environmental Policies. London: Cameron May LTD

Esty D (1999) Greening of the GATT (Trade, Environment and the Future) Washington D.C.. In: Institut für Umweltmanagement (IfU) (2001) Evaluation von Selbstverpflichtungen der Verbände der Chemischen Industrie. Unveröffentlichte Studie, Oestrich-Winkel

Etemad B, Luciani J, Bairoch P, Toutain JC (1991) World Energy Production 1800-1985. Geneva: Libriairie DROZ

European Commission, DG Energy & Transport (2002) Work Programme for SAVE and ALTERNER Calls 2001–2002, Brüssel

European Commission, UNDP (1999) Energy as a Tool for Sustainable Development for African, Caribbean and Pacific countries. New York

European Foundation (2000) The Role of the Social Partners in Sustainable Development. Conference Report. Dublin

E.V.A. [Energieverwertungsagentur] (2001) Informationen zum Bereich „nichtnukleare Energie" im 6. Rahmenprogramm für FTE der EU. Von Andreas Indinger, Wien. http://energytech.at/foerderung/6rp_index.html

E.V.A. (2000) Die Österreichische Energiepolitik im Hinblick auf erneuerbare Energiequellen. http://www.eva.wsr.ac.at/projekte/ren-in-a01.htm, 10/2/01

Farla JCM, Blok K (2001) Industrial long-Term Agreements on Energy Efficiency in the Netherlands. Paper for Journal of Cleaner Production

Faucheux S, Muir E, O'Conner M (1997) Neoclassical Natural Capital Theory and „Weak" Indicators for Sustainability. In: Land Economics, 73, p 528–552

Federal Energy Research and Development for the Challenges of the 21st Century (1997) President's Committee of Advisors on Science and Technology (PCAST). Washington D.C.

Fickl S, Raimund W (1999) Fuel economy labelling of cars and its impacts on buying behaviour, fuel efficiency and CO_2 reduction. Vortrag gehalten bei SAVE - For An Energy Efficient Millenium. The Conference. Session IV, Energy Efficient Equipment, 8.–10. November 1999, Graz

Fleischmann G (2001) Volkswirtschaftslehre. In: Ropohl G (ed) Erträge der interdisziplinären Forschung. Eine Bilanz nach 20 Jahren. Berlin: Schmidt, p 145–164

von Flotow P, Schmidt J (2001) Evaluation von Selbstverpflichtungen der Verbände der Chemischen Industrie. Arbeitspapiere des Instituts für Ökologie und Unternehmensführung e.V. (IÖU) Band 36, Oestrich-Winkel

von Flotow P, Steger U (2000) Die Brennstoffzelle – Ende des Verbrennungsmotors. Bern: Paul Haupt Verlag

Freeman C, Soete L (1997) The Economics of Industrial Innovation, Cambridge

Frenz W (2002) Warenverkehrsfreiheit und umweltschutzbezogene Energiepolitik. In: Natur und Recht, Heft 3, p 204 ff

Frenz W (2001a) Selbstverpflichtungen der Wirtschaft. Tübingen: Mohr Siebeck

Frenz W (2001b) Bergrecht und Nachhaltige Entwicklung. Berlin: Duncker & Humblot

Frenz W (2000), Sustainable Development durch Raumplanung. Berlin: Duncker & Humblot

Frenz W (1999a) Freiwillige Selbstverpflichtungen/Umweltvereinbarungen zur Reduzierung des Energieverbrauchs im Kontext des Gemeinschaftsrechts. In: EuR-Heft 1, p 27–48

Frenz W (1999b) Energiesteuern und Beihilfeverbot. In: EuZW, p 616 ff

Frenz W (1997) Nationalstaatlicher Umweltschutz und EG-Wettbewerbsfreiheit. Köln: Carl Heymanns Verlag

Frenz W, Unnerstall H (1999) Nachhaltige Entwicklung im Europarecht. Baden-Baden: Nomos

Fritsch M, Wein T, Ewers HJ (2001) Marktversagen und Wirtschaftspolitik. Mikroökonomische Grundlagen staatlichen Handelns. München: Vahlen

G8 Renewable Energy Task Force (2001) Final Report. July, 2001, www.renewabletaskforce.org

Gather C, Steger U (2001) Ökonomische und ökologische Auswirkungen der europäischen Deregulierung des Strommarktes. In: Hanekamp G, Steger U (ed) Nachhaltige Entwicklung und Innovation im Energiebereich. Graue Reihe Nr. 28, Europäische Akademie, Bad Neuenahr-Ahrweiler, p 116–136

GATT (1992) Trade and the environment. GATT-Report

Gerken L (1996) (ed) Ordnungspolitische Grundfragen einer Politik der Nachhaltigkeit. Baden-Baden: Nomos

Gethmann CF (1999) Rationale Technikfolgenbeurteilung. In Grunwald A (ed) Rationale Technikfolgenbeurteilung. Konzepte und methodische Grundlagen. Berlin: Springer

Gethmann CF, Kamp G (2000) Gradierung und Diskontierung von Verbindlichkeiten bei der Langzeitverpflichtung. In: Mittelstraß J (ed) Die Zukunft des Wissens. Berlin: Akademie Verlag. Auch in: Birnbacher D, Brudermüller G (2001) Zukunftsverantwortung und Generationensolidarität. Würzburg: Königshausen & Neumann, p 137–153

Goldemberg J, Johansson TB, Reddy AKN, Williams RH (1985) Basic Needs and Much More With One Kilowatt Per Capita. In: Ambio 14, p 190–200

Goodin RE (1982) Political Theory and Public Policy. Chicago: The University of Chicago Press

Goodin RE (1985) Protecting the Vulnerable. A Reanalysis of Our Social Responsibilities. Chicago: The University of Chicago Press

Goodin RE (1992) Green Political Theory. Cambridge: Polity Press

Goodin RE (1996) Enfranchising the Earth and its Alternatives. In: Political Studies, XLIV, p 835–849

Graafland JJ, Huizinga FH (1999) Taxes and Benefits in a Non-Linear Wage Equation. In: De Economist 147, No.1, p 39–54

Grübler A, Nakićenović N (1997) Decarbonizing the Global Energy System. IIASA Report RR-97-6, Laxenburg, Austria. Reprinted from Technology Forecasting and Social Change 53 (1996), p 97–110

Grunwald A (2000) Technik für die Gesellschaft von morgen. Möglichkeiten und Grenzen gesellschaftlicher Technikgestaltung. Frankfurt/Main: Campus

Haas H (1975) Technikfolgenabschätzung. München: Oldenbourg

Hall DO, Rosillo-Calle F, Williams RH, J. Woods J (1993) Biomass for energy – supply prospects. In: Th. Johansson et al.(ed) Renewable Energy - Sources for Fuels and Electricity. Island Press, Washington, D.C.

Hampicke U (1992): Ökologische Ökonomie. Individuum und Natur in der Neoklassik. Natur in der ökonomischen Theorie, Teil 4. Opladen: Westdeutscher Verlag.

Hampicke U (1999) Das Problem der Verteilung in der Neoklassischen und in der Ökologischen Ökonomie. In: Jahrbuch für Ökologische Ökonomik, Band 1. Marburg: Metropolis, p 153–188

Hanekamp G (2001) Scientific Policy Consulting and Participation. In: Poiesis & Praxis 1, p 78–84

Hanekamp G (2003) Decision theoretic arguments as heuristics in environmental policy decisions. In: Poiesis & Praxis 1, p 219–230 (published online Nov 2002)

Hartwick J (1977) Intergenerational Equity and the Investing of Rents from Exhaustible Resources. In: American Economic Review, Vol 67, p 972–974

Haupt U, Pfaffenberger W (2001) Wettbewerb auf dem deutschen Strommarkt – Drei Jahre nach der Liberalisierung. Beitrag für die 2. Internationale Energiewirtschaftstagung an der TU Wien

Von Hengel E (1998) Duurzaamheid: grenzen aan pluralisme. In: Filosofie & Praktijk, 19/3, p 113–127

Heuss E (1965) Allgemeine Markttheorie, Tübingen

Hicks JR (1932) The Theory of Wages. London: Macmillan

Hildebrandt E, Schmidt E (1997) Ökologisierung der Arbeit und die Innovationsfähigkeit der industriellen Beziehungen. In: Naschold F et al. (ed) Ökonomische Leistungsfähigkeit und institutionelle Innovation. Das deutsche Produktions- und Politikregime im globalen Wettbewerb. Berlin: Edition Sigma, p 183–210

Holmlund B, Kolm AS (2000) Environmental Tax Reform in a Small Open Economy With Structural Unemployment. In: International Tax and Public Finance, 7, p 315–333

Homann K (1996) Sustainability: Politikvorgabe oder regulative Idee? In: Gerken L (ed) Ordnungspolitische Grundfragen einer Politik der Nachhaltigkeit. Baden-Baden: Nomos, p 33–47

Hubbertpeak (2001). http://www.hubbertpeak.com.

Hübner H (2002) Nachhaltigkeit als Herausforderung für ganzheitliche Erneuerungsprozesse. Berlin: Erich Schmidt Verlag
http://www.ez.nl/beleid/home_ond/energiebesp/firstlevel_index.html, 9/27/01
http://www.ez.nl/Persberichten/persberichten2001/2001113.htm, 9/27/01
http://www.ez.nl/Persberichten/persberichten2001/2001128.htm, 9/27/01
http://www.senter.nl/energiebesparing/pb17092001.htm, 9/27/01
IEA (International Energy Agency) (2001a) Key World Energy Statistics – 2001 Edition. Paris
IEA (International Energy Agency) (2001b) Tagung des Verwaltungsrates auf Ministerebene, 15.–16. Mai. Kommunique, www.iea.org
IEA (International Energy Agency) (2000) The potential of wind energy to reduce CO_2-Emissions. Greenhouse Gas R&D Programme. Paris
IEA (International Energy Agency) (1998) Energy Technology Price Trends and Learning. Paris
IFOK Institut für Organisationskommunikation (ed) (1997) Bausteine für ein zukunftsfähiges Deutschland. Diskursprojekt im Auftrag von VCI und IG Chemie-Papier-Keramik. Wiesbaden
Imboden DM (2000) Energy forecasting and atmospheric CO_2 perspectives: Two worlds ignore each other. Integrated Assessment 1, p 321–330
Imboden D (1993) The Energy Needs of Today are the Prejudices of Tomorrow. In: GAIA 2, No 6, p 330–337
Imboden DM, Roggo C (2000) Die 2000 Watt-Gesellschaft – Der Mondflug des 21. Jahrhunderts. ETH Bulletin 276, p 24–27
Imboden DM, Jaeger CC (1999) Towards a Sustainable Energy Future. In: Energy – The Next Fifty Years. OECD, Paris
IMF, World Bank (1999) Poverty Reduction Strategies Papers. www.imf.org/external
Institute for Policy Studies (1997) The World Bank and the G7: Still Changing the Earth's Climate for Business. Washington, D.C.
Interdisciplinary Analysis of Successful Implementation of Energy Efficiency in the industrial, commercial and service sector (1998) Contract JOS3-CT95-0009, Final Report, Vol I. Wuppertal, Wien, Karlsruhe, Kiel, Copenhagen
IPCC (2001a) Climate Change 2001: The Scientific Basis. Cambridge
IPCC (2001b) Climate Change 2001: Impacts, Adaptation and Vulnerability. Cambridge
IPCC (2001c) Climate Change 2001: Mitigation. Cambridge
IPCC (2000). Special Report on Emissions Scenarios. A Special Report of Working Group III of the Intergovernmental Panel on Climate Change. Cambridge: Cambridge University Press
IPPC-Richtlinie (Integrated Pollution Prevention and Controll Directive) 96/61/EC, Brüssel
Jänicke M (2000) Ökologische Modernisierung als Innovation und Diffusion in Politik und Technik: Möglichkeiten und Grenzen des Konzeptes. Discussion Paper FFU-dp 1-2000, Berlin
Jänicke M, Mez L, Bechsgaard P, Klemmensen B (1998) Innovationswirkungen branchenbezogener Regulierungsmuster am Beispiel energiesparender Kühlschränke in Dänemark. Teilprojekt des Forschungssverbundes Innovative Auswirkungen umweltpolitischer Instrumente (FIU), Berlin
Johansson TB, Kelly H, Reddy AKN, Williams RH (ed) (1993) Renewable Energy – Sources for Fuels and Electricity. Island Press, Washington, D.C.
Keat R (1994) Citizens, Consumers and the Environment: Reflections on the "The Economy of the Earth". In: Environmental Values, Vol 3 No 3, p 333–349
Kemp R (2001) Possibilities for a Green Industrial Policy from an Evolutionary Technology Perspective. In: Binder M, Jänicke M, Petschow U (ed) Green Industrial Restructuring. Berlin: Springer, p 159–161
Kemp R (2000) Integrated Product Policy and Innovation: Incremental Steps and Their Limits. In: Ökologisches Wirtschaften, Nr. 6, p 24f
Kern K, Jörgens H, Jänicke M (2000) Die Diffusion umweltpolitischer Innovationen. Ein Beitrag zur Globalisierung von Umweltpolitik. ZfU, p 507–546
Klemmer P (1999) (ed) Innovationen und Umwelt. Berlin, Analytica
Klemmer P, Lehr U, Löbbe K (1999) Umweltinnovationen. Anreize und Hemmnisse. Berlin, Analytica
Knebel J, Wicke L, Michael G (1999) Selbstverpflichtungen und normsetzende Umweltverträge als Instrumente des Umweltschutzes (Umweltbundesamt: Berichte 99/5). Berlin: Erich Schmidt Verlag
Knoop S, Steger U (1998) Households: A new dimension of the IPPC-Directive? Unveröffentlichtes Manuskript, Oestrich-Winkel

KOM (2002) 43 endgültig (Kommission der Europäischen Gemeinschaften) Geänderte Vorschläge für Entscheidungen des Rates über die spezifischen Programme des 6. FTE-Rahmenprogramms. Brüssel

KOM (2001) 709 endgültig (Kommission der Europäischen Gemeinschaften) Geänderter Vorschlag für einen Beschluss des EP und des Rates über das 6. FTE-Rahmenprogramm. Brüssel

KOM (2001) 580 endgültig (Kommission der Europäischen Gemeinschaften) Mitteilung der Kommission über die Durchführung der ersten Phase des europäischen Programms zur Klimaänderung (ECCP). Brüssel

KOM (2001) 370 endgültig (Kommission der Europäischen Gemeinschaften) Weißbuch: Die europäische Verkehrspolitik bis 2010: Weichenstellungen für die Zukunft. Brüssel

KOM (2001) 279 endgültig (Kommission der Europäischen Gemeinschaften) Entscheidung des Rates über die spezifischen Programme zur Durchführung des 6. FTE-Rahmenprogramms. Brüssel

KOM (2001) 226 endgültig (Kommission der Europäischen Gemeinschaften) Vorschlag für eine Richtlinie des Europäischen Parlaments und des Rates über das Energieprofil von Gebäuden. Brüssel

KOM (2001) 68 endgültig (Kommission der Europäischen Gemeinschaften) Grünbuch zur integrierten Produktpolitik. Brüssel

KOM (2000) 769 endgültig (Kommission der Europäischen Gemeinschaften) Grünbuch. Hin zu einer europäischen Strategie für Energieversorgungssicherheit. Brüssel

KOM (2000) 247 endgültig (Kommission der Europäischen Gemeinschaften) Mitteilung der Kommission an den Rat, das Europäische Parlament, den Wirtschafts- und Sozialausschuss und den Ausschuss der Regionen. Aktionsplan zur Verbesserung der Energieeffizienz in der Europäischen Gemeinschaft. Brüssel

KOM (2000) 212 endgültig (Kommission der Europäischen Gemeinschaften) Mitteilung der Kommission an den Rat und das Europäische Parlament. Die Entwicklungspolitik der Europäischen Gemeinschaft. Brüssel

KOM (2000) 88 endgültig (Kommission der Europäischen Gemeinschaften) Mitteilung der Kommission an den Rat und das Europäische Parlament. Zu einem Europäischen Programm zur Klimaänderung (ECCP). Brüssel

KOM (2000) 31 endgültig (Kommission der Europäischen Gemeinschaften) Mitteilung der Kommission an den Rat, das Europäische Parlament, den Wirtschafts- und Sozialausschuss und den Ausschuss der Regionen zum 6. Umweltaktionsprogramm. Brüssel

KOM (2000) 6 endgültig (Kommission der Europäischen Gemeinschaften) Mitteilung der Kommission an den Rat, das Europäische Parlament, den Wirtschafts- und Sozialausschuss und den Ausschuss der Regionen. Hin zu einem europäischen Forschungsraum. Brüssel.

KOM (1996) 651 endgültig (Kommission der Europäischen Gemeinschaften) Mitteilung der Kommission an den Rat und das Europäische Parlament über Umweltvereinbarungen. Brüssel

KOM (1992) 23 (Kommission der Europäischen Gemeinschaften) Fünftes Umweltaktionsprogramm. Brüssel

Kong N, Salzmann O, Steger U (2002) Moving Corporations to promote Sustainable Consumption: The Role of NGOs. In press

Kopfmüller J, Brandl V, Jörissen J, Paetau M, Banse G, Coenen R, Grunwald A (2001) Nachhaltige Entwicklung integrativ betrachtet. Konstitutive Elemente, Regeln, Indikatoren. Berlin: Edition Sigma

Koskela E, Schöb R (1999) Alleviating unemployment: The case for green tax reforms. In: European Economic Review 43, p 1723–1746

Koskela E, Schöb R, Sinn HW (2001) Green Tax Reform and Competitiveness, Vol 2, Issue 1, February 2001, p 19–30

Krcal H-C (2000) Umweltschutzkooperationen in der Automobilindustrie – ein Überblick. In: UmweltWirtschaftsForum, Vol 8, No 2, p 5–10

Kurz R (1997) Unternehmen und nachhaltige Entwicklung. In: de Gijsel P et al. (ed) Ökonomie und Gesellschaft. Jahrbuch 14: Nachhaltigkeit in der ökonomischen Theorie. Frankfurt/Main: Campus, p 78–102

Kurz R (1996) Innovationen für eine zukunftsfähige Entwicklung. Aus Politik und Zeitgeschichte. In: Beilage zur Wochenzeitung Das Parlament B7/96, p 14–22

Kurz R, Graf HW, Zarth M (1989) Der Einfluss wirtschafts- und gesellschaftspolitischer Rahmenbedingungen auf das Innovationsverhalten von Unternehmen. Gutachten im Auftrag des Bundesministers für Wirtschaft, Tübingen

Langniß O (2000) Die Bedeutung grünen Stroms im liberalisierten Markt. Teil der Studie Klimaschutz durch erneuerbare Energien. Im Auftrag des Umweltbundesamtes und des Bundesministeriums für Umwelt, Naturschutz und Reaktorsicherheit, Stuttgart

Lapidus I, Looser U, Meier-Reinhold H, Müller-Groeling A, Paulse T, Vahlenkamp T (2000) Risiko-Management als Wettbewerbsvorteil im Strommarkt. In: EnergiewirtschaftlicheTagesfragen 9/2000, p 632–638

Lautenbach S, Steger U, Weihrauch P (1992) Evaluierung freiwilliger Branchenvereinbarungen im Umweltschutz. Freiwillige Kooperationslösungen im Umweltschutz. Ergebnisse eines Gutachtens und Workshops. Bundesverband der Deutschen Industrie e.V. (BDI-Drucksache, Nr. 249), Köln

Leitschuh-Fecht H (ed) (2001) Aktiv für die Zukunft – Wege zum nachhaltigen Konsum. UBA-Texte 37/01, Berlin

Leitschuh-Fecht H (2002) Lust auf Stadt. Ideen und Konzepte für urbane Mobilität. Bern: Paul Haupt

Lerch A, Nutzinger HG (2001) Nachhaltigkeit in wirtschaftsethischer Perspektive. In: Rissener Rundbrief 10-11, Oktober/November, p 61–79

Letchumanan R, Kodama F (2000) Reconciling the conflict between the ,pollution-haven' hypothesis and an emerging trajectory of international technology transfer. Research Policy 29, p 59–79

Lubbers R, Koorevaar J (2000) Primary Globalisation and the Sustainable Development Paradigm – Opposing Forces in the 21st Century. In: OECD (ed) The Creative Society of the 21st Century. Paris, p 173–189

Luiten E (2001) Beyond Energy Efficiency. Ph.D. Thesis, Utrecht University

Markusen JR, Morey ER, Olewiler NO (1993) Environmental policy when market structure and plant locations are endogenous. In: Journal of Environmental Economics and Management, Vol 35, p 69–86

Martinot E, McDoom O (2000) Promoting Energy Efficiency and Renewable Energy. GEF Climate Change Projects and Impacts. Global Environment Facility, Washington, D.C.

Matthews R (1949-50) Reciprocal Demand and Increasing Returns. In: Review of Economic Studies, Vol 17, No 42, p 149–158

McGuire MC (1982) Regulation, Factor Rewards, and International Trade In: Journal of Public Economics 17, p 335–354

Merrifield JD (1988) The Impact of Selected Abatement Strategies on Transnational Pollution. The Terms of Trade and Factor Rewards: A General Equilibrium Approach. In: Journal of Environmental Economics and Management, Vol 29, p 259–284

Möschel W (1994) Innovationspolitik als Ordnungspolitik. In: Ott C, Schäfer HB (eds) Ökonomische Analyse der rechtlichen Organisation von Innovationen. Tübingen: Mohr, p 40–58

de Moor APG, van Beers CP (2001) Het Internationale klimaatcompromis. In: Economisch-Statistische Berichten, p 552–554

Nakićenović N, Grübler A, McDonald A (1998) Global Energy Perspectives. Cambridge: Cambridge University Press

Newell R, Jaffe AB, Stavins RN (1999) The Induced Innovation Hypothesis and Energy-Saving Technological Change. In: The Quarterly Journal of Economics, Vol. 114 (3), p 941–976

Nielsen SB, Pedersen LH, Sørensen PB (1995) Environmental Policy, Pollution, Unemploymnet, and Endogenous Growth. In: International Tax and Public Finance, 2(2), p 185–205

Nill J, Petschow U, Jahnke M (2001) New Theoretical Perspectives on Industrial Restructuring and their Implications for (Green) Industrial Policy. In: Binder M, Jänicke M, Petschow U (eds) Green Industrial Policy. International Case Studies and Theoretical Interpretations. Berlin: Springer, p 73–96

Nitsch J, Rösch C (2001) Perspektiven für die Nutzung regenerativer Energien. in: Grunwald A, Coenen R, Nitsch J, Sydow A, Wiedemann P (eds) Forschungswerkstatt Nachhaltigkeit. Berlin: Edition Sigma, p 291–324

Nutzinger HG, Radke V (1995a) Das Konzept der nachhaltigen Wirtschaftsweise. In: Nutzinger H (ed) Nachhaltige Wirtschaftsweise und Energieversorgung. Konzepte, Bedingungen, Ansatzpunkte. Marburg: Metropolis, p 13–49

Nutzinger HG, Radke V (1995b) Wege zur Nachhaltigkeit. In: Nutzinger H (ed) Nachhaltige Wirtschaftsweise und Energieversorgung. Konzepte, Bedingungen, Ansatzpunkte. Marburg: Metropolis, p 225–256

OECD (2001) Policies to Enhance Sustainable Development. Paris

OECD/IEA (2000a) Energy Labels & Standards. Paris
OECD/IEA (2000b) Experience Curves for Energy Technology Policy. Paris
OECD (1999) Voluntary Approaches for environmental policy. An assessment. Paris
OECD (1996) Shaping the 21st Century: The Contribution of Development Co-operation. Paris
OECD (1995) Global Warming: Economic Dimension and Policy Responses, Paris
Onigkeit J, Alcamo J (2000) Stabilisierungsziele für Treibhausgaskonzentrationen. Eine Abschätzung der Auswirkungen und der Entwicklungspfade. Universität Kassel.
Ossebaard ME, van Wijk AJM, van Wees MT (1997) Heat Supply in the Netherlands: A Systems Analysis of Costs, Exergy Efficiency, CO_2 and NO_x Emissions. In: Energy 22, p 1087–1098
o.V. (2000) Produktkampagne Top 10 – Eine ungewöhnliche Kooperation. In: Öko-Institut (ed) Öko-Mitteilungen. Ausgabe 3-4/00. Freiburg, p 25
Pearce D, Atkinson G (1993) Measuring Sustainable Development. In: Ecodecision, June 1993, p 64–66
Petersen T, Faber M, Schiller J (2000) Umweltpolitik in einer evolutionären Wirtschaft und die Bedeutung einer institutionellen Umweltökonomik. In: Bizer K, Linscheidt B, Truger A (eds) (2000) Staatshandeln im Umweltschutz. Perspektiven einer institutionellen Umweltökonomik. Berlin, p 135–150
Pethig R (1976) Pollution, Welfare, and Environmental Policy in the Theory of Comparative Advantage. In: Journal of Environmental Economics and Management 2, p 160–169
Pimm S (2001) The World According to Pimm, McGraw-Hill, p 304 ff
Pissarides C (1998) The impact of employment tax cuts on unemployment and wages. The role of unemployment benefits and tax structure. European Economic Review 42, p 155–183
Popp D (2001) Induced Innovation and Energy Prices. NBER Working Paper No 8284, Cambridge MA
Projektgruppe Mobilität (2001) Kurswechsel im öffentlichen Verkehr. Mit automobilen Angeboten in den Wettbewerb. Berlin: Edition Sigma
Radgen, P, Jochem E (eds) (1999) Energie effizient nutzen – Schwerpunkt Strom. Frauenhofer Institut für Systemtechnik und Innovationsforschung (ISI) Karlsruhe
http://www.baden-wuerttemberg.de/sixcms_upload/media/110/stromsparinitiative_modellprojekte_ und_fachartikel.pdf
Rawls J (1971/1998) A Theory of Justice (revised edition 1998). Berlin: Akad.-Verlag
Rehbinder E (1991) Das Vorsorgeprinzip im internationalen Vergleich. Düsseldorf: Werner-Verlag
Rehbinder E (1998) Ziele, Grundsätze, Strategien, Instrumente. In: Salzwedel J et al. (eds) Grundzüge des Umweltrechts. Berlin: Schmidt
Renn O (1984) Risikowahrnehmung der Kernenergie. Frankfurt/Main: Campus
Rennings K (ed) (1999) Innovation durch Umweltpolitik. Baden-Baden
Rennings K, Brockmann KL, Bergmann H (1997) Voluntary Agreements in Environmental Protection – Experiences in Germany and future Perspectives. ZEW Discussion Paper No. 97-04 E, Berlin
Rennings K, Brockmann KL, Bergmann H, Kühn I (1996) Nachhaltigkeit, Ordnungspolitik und freiwillige Selbstverpflichtungen. ZEW Schriftenreihe, Umwelt- und Ressourcenökonomie, Heidelberg
Requate T (1998) Incentives to Innovate under Emission Taxes and Tradeable Permits. European Journal of Political Economy 14, No 1, p 139–165
Rietbergen MG, Farla JCM und Blok K (2001) Do Agreements Enhance Energy Efficiency Improvement? Journal of Cleaner Production 10 (2002) p. 153–163
Ritter H, Amann E (2001) Energy+: Kühle Europameister für kühle Rechner. In: Energieverwertungsagentur – the Austrian Energy Agency (E.V.A.): energy. Die Zeitschrift der Energieverwertungsagentur, No 2/2001, p 26–28
Rogner HH (1997) An Assessment of World Hydrocarbon Resources. In: Annual Review of Energy and Environment, Vol 22, p 217–262
Rosenberg N (1982) Inside the Black Box: Technology and Economics. New York
Der Sachverständigenrat für Umweltfragen (SRU) (1994) Umweltgutachten 1994. Für eine dauerhaft umweltgerechte Entwicklung. Stuttgart: Metzler-Poeschel
Der Sachverständigenrat für Umweltfragen (SRU) (1996) Umweltgutachten 1996: Zur Umsetzung einer dauerhaften umweltgerechten Entwicklung. Stuttgart: Metzler-Poeschel
Der Sachverständigenrat für Umweltfragen (SRU) (2000) Umweltgutachten 2000: Schritte ins nächste Jahrtausend. Stuttgart: Metzler-Poeschel

Schindler J, Zittel W (2000) Öffentliche Anhörung von Sachverständigen durch die Enquête Kommission des Deutschen Bundestages „Nachhaltige Energieversorgung unter den Bedingungen der Globalisierung und der Liberalisierung" zum Thema „Weltweite Entwicklung der Energienachfrage und der Ressourcenverfügbarkeit". Schriftliche Stellungnahme zu ausgewählten Fragen der Kommission, Ottobrunn

Schlegelmichel K (2000) Energy Taxation in the EU – Recent Processes. Heinrich-Böll Stiftung, Brüssel

Schlesinger M, Schulz W Deutscher Energiemarkt 2020. Prognose im Zeichen von Umwelt und Wettbewerb. In: Energiewirtschaftliche Tagesfragen, 3/2000, p 106–113

Schlomann B, Eichhammer W, Gruber E, Kling N, Mannsbart W (Fraunhofer ISI), Stöckle F (GfK Marketing Services) (2001) Evaluierung zur Umsetzung der Energieverbrauchskennzeichnungsverordnung (EnVKV). Projektnummer 28/00. Kurzfassung des Abschlussberichts an das Bundesministerium für Wirtschaft und Technologie. Karlsruhe

Schmitt D, Düngen H (1995) Energie und Umwelt. In: Junkernheinrich M, Klemmer P, Wagner GR (eds) Handbuch zur Umweltökonomie. Berlin: Analytica, p 22–26

Schneider K (1997) Involuntary Unemployment and Environmental Policy: The Double Dividend Hypothesis. Scandinavian-Journal-of-Economics, Vol 99(1), p 45–49

Scholz CM (1998) Involuntary Unemployment and environmental policy: The Double Dividend Hypothesis: A Comment. In: Scandinavian Journal of Economics, Vol 100 (3), p 663–664

Schröder M, Claussen M, Grunwald A, Hense A, Klepper G, Lingner S, Ott K, Schmitt D, Sprinz D (2002) Klimavorhersage und Klimavorsorge. Berlin: Springer

Schumpeter JA (1911) Theorie der wirtschaftlichen Entwicklung. Eine Untersuchung über Unternehmergewinn, Kapital, Kredit, Zins und den Konjunkturzyklus, 6. Auflage 1964. Berlin: Duncker & Humblot

Schumpeter JA. (1942) Capitalism, Socialism and Democracy. New York

SEK (2002) 105 endgültig Mitteilung der Kommission an das Europäische Parlament, Brüssel

Sekretariat der Klimarahmenkonvention (ed) (1992) Klimarahmenkonvention (KRK). New York (ebenfalls abgedruckt in BGBl. II 1993, S 1784 ff)

Sen A (1987) On Ethics and Economics. Oxford: Blackwell

Shreeve S, von Flotow P (2001) Sustainable Consumption and the Internet. Unpublished working paper for IMD's Forum for Corporate Sustainability, Lausanne

Sinclair P (1992) High does nothing and rising is worse: Carbon taxes should keep declining to cut harmful emissions. The Manchester School, Vol LX No.1, March, p 41–52

Soete LLG, Ziesemer T (1997) Gains from Trade and Environmental Policy under Imperfect Competition and Pollution from Transport. In: Feser HD, Hauff M (eds) Neuere Entwicklungen in der Umweltökonomie und -politik. Volkswirtschaftliche Schriften Universität Kaiserslautern. Regensburg: Transfer Verlag, p 249–268

Sohmen E (1976) Allokationstheorie und Wirtschaftspolitik. Tübingen: Mohr

Solow, RM (1974) The Economics of Resources or the Resources of Economics. In American Economic Review, LXIV (2), p 1–14

Staudt E, Kottmann M, Meier AJ (2001) Kompetenzverfügbarkeit und innovationsorientierte Regionalentwicklung. In: List Forum 27, p 346–364

Steger U (2001) Globalisierung, Nachhaltigkeit und Elitenkooperation. In: Müller-von-Maiborn B, Steger U (eds) Elitenkooperation in der Region. Essen, p 15–30

Steger U (2000) Environmental Management Systems: Empirical Evidence and further Perspectives. In: European Management Journal, Vol 18, p 22–37

Steger U (1998) The Strategic Dimension of Environmental Management. London

Strand J (1996) Environmental policy, worker moral hazard, and the double dividend issue. In: CarraroC, Siniscalco D (eds) Environmental Fiscal Reform and Unemployment. Dordrecht: Kluwer, p 121–135

Streffer C, Bücker J, Cansier A, Cansier D, Gethmann CF, Guderian R, Hanekamp G, Henschler D, Pöch G, Rehbinder E, Renn O, Slesina M ,Wuttke K (2000) Umweltstandards. Kombinierte Expositionen und ihre Auswirkungen auf den Menschen und seine Umwelt. Berlin: Springer

Sustainable Energy & Economic Network (SEEN) (2001) New Database Calculates Lifetime Greenhouse Gas Emissions from Nine Years of World Bank Fossil Fuel Projects. Press release October 29, 2001, www. seen.org

Swisher J, Wilson D (1993) Renewable energy potentials. In: Energy, Vol 18, p 437–459

TAB (Büro für Technikfolgenabschätzung beim Deutschen Bundestag) (2000) Arbeitsbericht 69: Elemente einer Strategie für eine nachhaltige Energieversorgung. Vorstudie. TAB, Berlin

Turner RK, Doktor P, Adger N (1994) Sea-Level Rise & Costal Wetlands in the U.K.: Mitigation Strategies for Sustainable Management In: Jansson AM et al. (ed) Investing in Natural Capital. The Ecological Economics Approach to Sustainability. Washington D.C.: Island Press, p 266–290

Tweede Kamer der Staten-Generaal (2001) Interdepartementaal Beleidsonderzoek: Energiesubsidies. Brief van de Minister van Economische Zaken. Vergaderjaar 2001-2002, Nr.1 und 2, Sdu Uitgevers, 's-Gravenhage

UNDP/OECD/WEC (2000) World Energy Assessment. Energy and the challenge of sustainability. New York

UNDP/UNDESA/WEC (2000) World Energy Assessment. United Nations Development Programme. New York

UNFCCC (2001) Review of the Implementation of Commitments and of other Provisions of the convention. Fccc/CP/2001/L.7, 24 July

United States Senate (1972) Technology Assessment Act of 1972. Report of the Committee on Rules and Administration. Washington D.C.

USGS (2000). World Petroleum Assessment 2000. United States Geological Survey (USGS), Washington D.C.

Vallance E (1995) Business ethics at work. Cambridge

Vermeend W, van der Vaart J (1997) Greening Taxes: The Dutch Model. Paper for the European Association of Environmental and Resource Economics (EAERE). Eight Annual Conference, Tilburg, The Netherlands, 26-28 June, 1997

Wagner H (1989) Stabilitätspolitik. München: Oldenbourg

Waide P (1999) Market analysis and effect of EU labelling and standards: The example of cold appliance. Vortrag gehalten bei SAVE - For An Energy Efficient Millenium. The Conference. Session IV, Energy Efficient Equipment, 8.–10. November 1999, Graz

WBGU (Wissenschaftlicher Beirat der Bundesregierung Globale Umweltveränderungen) (2001) Die Chance von Johannesburg - Eckpunkte einer Verhandlungsstrategie. Berlin

WBGU (Wissenschaftlicher Beirat der Bundesregierung Globale Umweltveränderungen) (1997) Ziele für den Klimaschutz. Stellungnahme zur dritten Vertragsstaatenkonferenz der Klimarahmenkonvention in Kyoto. Bremerhaven

WBGU (Wissenschaftlicher Beirat der Bundesregierung Globale Umweltveränderungen) (1996) Welt im Wandel. Berlin: Springer

WBGU (Wissenschaftlicher Beirat der Bundesregierung Globale Umweltveränderungen) (1995) Globale Szenarien zur Ableitung globaler CO_2-Reduktionsziele und Umsetzungsstrategien. Stellungnahme zur ersten Vertragsstaatenkonferenz der Klimarahmenkonvention. Bremerhaven

WCED (Weltkommission für Umwelt und Entwicklung) (1987) Unsere gemeinsame Zukunft. Greven: Eggenkamp (engl. Original: Our Common Future. Oxford 1987)

WEC (2001) Survey of Energy Resources 1998. World Energy Council

Weegink RJ (1998) Basisonderzoek Aardgas Kleinverbruikers BAK 1997, EnergieNed, Arnhem

Wegner, G. (2001) Marktkonforme Wirtschaftspolitik und evolutorische Ökonomik. Vortragsmanuskript zur Tagung des Ausschusses für Evolutorische Ökonomik beim Verein für Socialpolitik in St. Gallen.

Wegner G (1991) Wohlfahrtsaspekte evolutorischen Marktgeschehens: neoklassisches Fortschrittsverständnis und Innovationspolitik aus ordnungstheoretischer Sicht. Tübingen: Mohr

Weiss MA, Heywood JB, Drake EM, Schafer A, AuYeung FF (2000) On the Road in 2020. Energy Laboratory, MIT, Cambridge, MA

von Weizsäcker E, Lovins AB, Lovins LH (1995) Faktor Vier. München: Droemer Knauer

von Westphalen R (1997) Technikfolgenabschätzung als politische Aufgabe. München: Oldenbourg

Wietschel M, Dreher M, Huber Th, Rentz O (2001) Grüne Angebote in Deutschland: Stand und Perspektiven. Vortrag gehalten anlässlich der 2. Internationalen Energiewirtschaftstagung IEWT 2001 vom 21.–23.2.2001, TU-Wien

Williams RH (1994) Die Renaissance der Energieindustrie. In: Steger U/Hüttl A (ed) Strom oder Askese? Auf dem Weg zu einer nachhaltigen Strom- und Energieversorgung. Frankfurt am Main: Campus, p 141–198

World Bank (2001) Making Sustainable Commitments. An Environmental Strategy for the World Bank. Washington, D.C.

World Bank (2000) Interim Report of the Implementation of the Fuel for Thought Strategy: An Environmental Strategy for the Energy Sector. Washington D.C.
World Bank (1999a) Fuel for Thought. An Environmental Strategy for the Energy Sector. Washington D.C.
World Bank (1999b) Comprehensive Development Framework. Washington D.C.
World Bank (1996) Monitoring environmental progress: Expanding the measure of wealth, Environment Department, Conference Draft. Washington D.C.
World Bank (1995) Monitoring environmental progress: A report on work in progress. Washington D.C.
Wortmann K (2000) Energieeffizienz im liberalisierten Markt. In: Energiewirtschaftliche Tagesfragen, Vol 50(6), p 438–443
WRI (1997) World Resources 1996-1997. World Resources Institute
WRI (2001) World Resources 2000-2001. World Resources Institute
Wuest & Partner (1999) Immo-Monitoring 2000. Bd. 3 Baumarkt, Zürich
Wüstenhagen R (2000) Ökostrom – von der Nische zum Massenmarkt. Entwicklungsperspektiven und Marketingstrategien für eine zukunftsfähige Elektrizitätsbranche. Zürich: vdf
WTO (2000) Trade and Environment in the WTO. 6 Seiten. http://www.wto.org/wto/environ/environ1.htm, download 29-3-00
Xu X (1999) Do Stringent environmental Regulations Reduce the International competitiveness of environmentally Sensitive Goods? World Development, Vol 27(7), p 1215–1226
Xu X, Song L (2000) Regional Cooperation and the environment: Do „Dirty" Industries Migrate? Weltwirtschaftliches Archiv, Vol 136(1), p 137–57
trag des Bundesministeriums für Bildung und Forschung, Bundestag-Drucksache 14/2057

List of abbreviations

Art	Article
BAT	Best available technology
GDP	Gross domestic product
BMBF	*Bundesministerium für Bildung und Forschung*[1]
BREF's	BAT reference documents
CDM	Clean Development Mechanism
CHP	Combined heat and power generation
CO_2	Carbon dioxide
COP's	Conferences of the parties
c.p.	ceteris paribus
ECCP	European Climate Change Programme
EC	European Community
EU	European Union
ECJ	European Court of Justice
FAO	Food and Agriculture Organization of the United Nations
GATT	General Agreement on Tariffs and Trade
GEF	Global Environment Facility
GGE	Greenhouse gas emissions
GW	Gigawatt
IEA	International Energy Agency
IIASA	International Institute for Applied Systems Analysis
IPCC	Intergovernmental Panel on Climate Change
IPPC	Integrated Pollution Prevention and Control Directive
ICT	Information and communication technology
LCA	Life cycle assessment
NGO	Non-governmental organization
OECD	Organization for Economic Co-operation and Development
R&D	Research and development
RTD	Research and technological development
SME	Small and medium-sized enterprises
SRU	*Sachverständigenrat für Umweltfragen*[2]
TW	Terawatt
UNDP	United Nations Development Programme
UNEP	United Nations Environment Programme
UNESCO	United Nations Educational, Scientific and Cultural Organization
UNIDO	United Nations Industrial Development Organization
UNFCCC	United Nations Framework Convention on Climate Change
UWG	*Gesetz gegen unlauteren Wettbewerb*[3]
WBGU	*Wissenschaftlicher Beirat Globale Umweltveränderungen*[4]
WCED	World Commission on Environment and Development
WEC	World Energy Council
WRI	World Resources Institute
WTO	World Trade Organization

---

[1] Federal Ministry for Education and Research (Germany).
[2] Council of experts on the environment (Germany).
[3] German law against unfair competition.
[4] Scientific advisory council on global changes of the environment (Germany).

List of Authors

Achterberg, Wouter, Dr. phil. † 16. June 2002. Studied philosophy at Amsterdam University. Lecturer in ethics and political philosophy at Amsterdam University from 1972. 1986: Doctoral thesis "Partners in de natuur". 1991–2002: "Sokrates" Professor (sponsored by the humanistic foundation of the same name) at the University of Wageningen. Main research interests: Ethics and political philosophy, especially the ethical and metaphysical foundations of an environmental philosophy.

Blok, Kornelis, Professor Dr., studied physics at Utrecht. Doctorate 1991, thesis: 'On the Reduction of Carbon Dioxide Emissions'. Professor of "Science and Society" at Utrecht University. Director of Ecofys, the energy and environmental consultancy. Main body of work on the: Development of energy technology, technology and politics. Lead author at the Intergovernmental Panel on Climate Change (IPCC). Address: University of Utrecht, Vakgroep Natuurwetenschap en Samenleving, Padualaan 14, 3584 CH Utrecht, The Netherlands.

Bode, Henning, studied law at Bonn University. 1999–2001: Research Associate with the teaching and research department for mining law and environmental law at RWTH Aachen under Professor Dr. Walter Frenz (see below). Main research areas: Regional planning law, energy law and soil protection law. Address: Hauptstraße 11, D-56412 Girod.

Frenz, Walter, Professor Dr. jur., born 1965, studied law and politics at Würzburg, Caen and Munich; 1994–1996: Research assistant at the University of Münster and professor of German (public) law at the university of Nijmegen; since 1997 Professor of mining law and environmental law at RWTH Aachen; major work also in the areas of energy law, European law and fiscal law. Address: Lehr- und Forschungsgebiet Berg- und Umweltrecht der RWTH Aachen, Wüllnerstraße 2, D-52062 Aachen.

Gather, Corinna, Dipl.-Vwl., studied at the University of Cologne and the FU Berlin; 1986–1989: Statistisches Landesamt Berlin; 1989–2000: Staff and board member at the environmental consultancies *Umweltberatungsstelle e.V.* and *KOFIRM e.V.* in Berlin; since 2000: Research work at the chair of Prof. Steger at the TU Berlin. Main interests: Deregulation of the energy markets, improving the energy efficiency in private households. Address: Institut für Technologie und Management, Hardenbergstr. 4-5, D-10623 Berlin.

Hanekamp, Gerd, Dr. phil. Dipl.-Chem., studied chemistry at Heidelberg and Marburg and at the École Nationale de Chimie in Lille; 1996: Doctorate in philosophy at the University of Marburg; since 1996: staff scientist with the European Academy Bad Neuenahr-Ahrweiler GmbH; fields of research: philosophy of science, linguistic philosophy, culturalistic theory of social sciences, the theory of technology assessment, business ethics.

Imboden, Dieter M., Professor Dr. sc. nat., Dipl. Phys., studied physics at Berlin, Basel and Zurich; since 1988: Professor for environmental physics at the Eidg. Technischen Hochschule in Zurich. Fields of research: Aquatic physics, modeling of environment systems, sustainable energy systems. Address: Professur Umweltphysik, ETH Zentrum VOD, Voltastrasse 65, CH-8092 Zürich.

Kost, Michael, Dipl. Umwelt-Natw. ETH, studied environmental sciences; doctoral student at the Professur Umweltphysik of the Eidgenössische Technische Hochschule (ETH) Zurich under Professor Dieter Imboden; thesis title (provisional): "Sustainable development of the *Constructed Switzerland*". Address: Professur Umweltphysik, ETH Zentrum VOD, Voltastrasse 65, CH-8092 Zürich.

Kurz, Rudi, Professor Dr. rer. pol., studies of economics and doctorate at Tübingen. 1978-1988: Research fellow at the *Institute for Applied Economic Research (IAW)*, Tübingen. Since 1988: Professor for economics at the Pforzheim University of Applied Sciences. Fields of work: Innovation research and environmental economics. Address: Hochschule Pforzheim FB 7, Tiefenbronner Str. 65, D-75175 Pforzheim.

Jahnke, Matthias, Dipl.-economist, studied economics at the University of Kassel. Research assistant and doctoral student at the "Theory of public and private enterprises" chair at the University of Kassel. Research interests: Environmental economics, energy economics, ecological economics. Address: University of Kassel, Fachbereich Wirtschaftswissenschaften, Nora-Platiel-Straße 4, D-34109 Kassel.

Nutzinger, Hans G., Prof. Dr. rer. pol., Dipl.-Volksw., Studies, doctorate and habilitation at Heidelberg; since 1978: Professor for the theory of public and private enterprises at the University of Kassel. Areas of work: Macroeconomic theory of enterprise, fundamental issues of economic policy, the history of dogmas, the ethics of the economy and enterprises, environmental economics and ecological economics. Address: University of Kassel, Fachbereich Wirtschaftswissenschaften, Nora-Platiel-Straße 4, D-34109 Kassel.

Steger, Ulrich, Prof. Dr. rer.pol., Dipl.-economist. Following three years of military service in Germany, studies and doctorate at Münster and Bochum; 1976: elected to the German *Bundestag* (National Parliament), 1984–87: Economy and Technology Minister in the federal state of Hessen; 1987–94: Professorship at the European Business School, Oestrich-Winkel; 1991–93: Volkswagen board member; 1995–99: Head of the *Forschungskolleg Globalisierung* of the Gottlieb Daimler and Karl Benz Foundation; since 1995: Alcan Professor for Environmental Management, IMD, Lausanne. Address: P.O. Box 915, CH 1001 Lausanne.

Ziesemer, Thomas, Dr. rer.pol., studied national economics at the universities of Kiel and Regensburg. Doctorate on the theory of underdevelopment. Qualified for professorship with a monograph on the causes of debt crises. Fields of research: International economic relations, economics of development, the environment and labor, growth and technological changes. Address: University of Maastricht, MERIT, P.O.Box 616, Tongersestraat 49, NL 6200 MD Maastricht.